I0020262

Hack the Cybersecurity Interview

Second Edition

Navigate Cybersecurity Interviews with Confidence, from Entry-level to Expert roles

Christophe Foulon

Ken Underhill

Tia Hopkins

Hack the Cybersecurity Interview
Second Edition

Senior Publishing Product Manager: Reshma Raman
Acquisition Editor – Peer Reviews: Gaurav Gavas
Project Editor: Amisha Vathare
Content Development Editor: Soham Amburle
Copy Editor: Safis Editing
Technical Editor: Simanta Rajbangshi
Proofreader: Safis Editing
Indexer: Rekha Nair
Presentation Designer: Ganesh Bhadwalkar
Developer Relations Marketing Executive: Meghal Patel

First published: July 2022
Second edition: August 2024

Production reference: 1060824

Published by Packt Publishing Ltd.
Grosvenor House

11 St Paul's Square
Birmingham
B3 1RB, UK.

ISBN 978-1-83546-129-7

www.packt.com

Contributors

About the authors

Christophe Foulon, founder and cybersecurity executive advisor at CPF Coaching LLC focuses on helping small to mid-sized businesses improve their security maturity, and grow their business in the process. He brings over 17 years of experience as a vCISO, information security manager, adjunct professor, author, and cybersecurity strategist, and a passion for customer service, process improvement, and information security. He has also spent over 10 years leading, coaching, and mentoring people. He gives back by writing books like this and *Develop Your Cybersecurity Career Path* and supporting non-profits like Whole Cyber Human Initiative and his local Rotary Club.

I would like to thank my wife and son, who patiently and lovingly support me as I find time to continue to write among all the other things we need to do. Without a support system like that, I would not have the bandwidth to help others like I do.

I would also like to thank Ken and Tia, as they have been such amazing co-authors and helped to pull together all this knowledge for our readers. It's an amazing feeling once it's completed.

Ken Underhill is an experienced cybersecurity executive and has helped over 2 million people build their cybersecurity skills. He has won multiple awards, including the Cyber Champion award, Best Cybersecurity Marketer, SC Media Outstanding Educator, and a 40 under 40 award. Ken volunteers with organizations like Minorities in Cybersecurity, Black Girls Hack, BBWIC, EKC, and sits on the global ethical hacking advisory board for EC-Council.

I would like to thank my spouse for her support as I make the world a better place. To all our readers, you already have everything you need within you to live the life of your dreams. Thank you to my co-authors and of course thank you to the amazing Mari Galloway for her contributions to both this book and my life.

Tia Hopkins is a global award-winning cybersecurity executive with over two decades of experience in IT and IT security. In addition to her primary role, she is an adjunct professor of cybersecurity, a women's tackle football coach, a keynote speaker, and a LinkedIn learning instructor.

Tia's extensive educational background includes a BS in information technology, an ms in information security and assurance, an ms in cybersecurity and information assurance, and an Executive Master of Business Administration. She holds industry certifications such as CISSP, CISM, and GSLC and is pursuing a PhD in cybersecurity leadership. Her research focuses on leveraging cyber resilience to bridge communication gaps between digital leaders, non-technical business leaders, and board directors.

Throughout her career, Tia has received numerous accolades, including a Lifetime Achievement Award from AmeriCorps and the Office of the President of the United States for her volunteer work focused on diversifying the cybersecurity talent pool. She also earned the SANS Difference Makers People's Choice Award for Team Leader of the Year.

Tia is a member of the Forbes Technology Council and has been featured in prominent publications like the Wall Street Journal, Dark Reading, and InformationWeek. She contributed a chapter to the book *The Rise of Cyber Women: Volume 2* and co-authored two best-selling books: *Hack the Cybersecurity Interview* and *Securing Our Future: Embracing the Resilience and Brilliance of Black Women in Cyber.*

Committed to diversity and inclusion, Tia founded Empow(H)er Cybersecurity, a non-profit organization that inspires and empowers women of color to pursue cybersecurity careers. She also serves on the board of Cyversity, a non-profit association dedicated to diversifying, educating, and empowering women, traditionally underrepresented minorities, and veterans in their cybersecurity careers.

I would like to thank all the amazing people who have supported me along the way. First off, to my fiancé - thank you for your endless love, patience, support, and encouragement. I couldn't have done this without you.

A huge shoutout to my co-authors, Ken and Chris, whose collaboration has been incredible. Your dedication and passion for cybersecurity have made this journey a great one, and I'm grateful for your partnership.

To everyone who read the first edition, your feedback and success stories have been incredibly inspiring. Knowing that this book has helped you on your career journey fuels my passion for writing and teaching. I hope this new edition continues to be a valuable resource for you.

About the reviewer

Derek Fisher offers over 25 years of experience in hardware, software, and cybersecurity, spanning industries like healthcare and finance. An accomplished leader and educator, he excels in cybersecurity strategy, risk management, and compliance, leading incident response efforts and directing high-performing teams. Derek effectively communicates complex technical concepts to a range of audiences, including executives and board members. In academia, he translates his professional knowledge into courses for both graduate and undergraduate students, and has developed self-paced online training programs on topics such as threat modeling and application security. Additionally, Derek is an award-winning author of a children's book series on online safety, recognized by the Mom's Choice Award, and has published a well-received guide on building application security programs.

Derek is the founder of Securely Built, which is dedicated to providing security services and education to individuals and businesses. You can find more here: `https://securelybuilt.com/`. He has also written the *Application Security Program Handbook* as well as a children's book series on cybersecurity called *Alicia Connected*.

Join us on Discord!

Read this book alongside other users. Ask questions, provide solutions to other readers, and much more.

Scan the QR code or visit the link to join the community.

https://packt.link/SecNet

Table of Contents

Chapter 2: Cybersecurity Engineer   19

Chapter 3: SOC Analyst   29

Chapter 4: Penetration Tester   53

Chapter 15: Behavioral Interview Questions   **239**

Chapter 16: Final Thoughts 287

Other Books You May Enjoy 299

Index 303

Preface

This book covers best practices for preparing yourself for cybersecurity job interviews. Most of the chapters cover a specific cybersecurity job and interview questions that you might be asked in the interview. The behavioral interview questions chapter covers questions the authors have been asked during interviews, regardless of job role. The final chapter of the book covers additional information from the authors on how to best prepare yourself for job interviews. Remember, your resume and networking with people can get you the interview, but preparation for the interview is what helps you get the job.

Who this book is for

This book is valuable to aspiring cybersecurity professionals looking to gain insight into the types of questions they might face during an interview. It's also for experienced cybersecurity professionals looking to level up their interview game.

What this book covers

Chapter 1, *Hacking Yourself*, is where you are going to learn about tips for preparing for your job interview, some general interview questions you might be asked, and how you should respond to them, and ways to handle stress.

Chapter 2, *Cybersecurity Engineer*, is where you will learn about the Cybersecurity Engineer career path and some of the common interview questions that are asked.

Chapter 3, *SOC Analyst*, is where you will learn about the SOC Analyst career path and some of the common interview questions that are asked.

Chapter 4, *Penetration Tester*, is where you will learn about the Penetration Tester career path and some of the common interview questions that are asked.

Chapter 5, *Digital Forensics Analyst*, is where you will learn about the digital forensics analyst career path and some of the common interview questions that are asked.

Chapter 6, *Cryptographer/Cryptanalyst*, is where you will learn about the cryptographer career path and some of the common interview questions that are asked.

Chapter 7, *GRC/Privacy Analyst*, is where you will learn about the GRC analyst career path and some of the common interview questions that are asked.

Chapter 8, *Security Auditor*, is where you will learn about the security auditor career path and some of the common interview questions that are asked.

Chapter 9, *Malware Analyst*, is where you will learn about the malware analyst career path and some of the common interview questions that are asked.

Chapter 10, *Cybersecurity Manager*, is where you will learn about the cybersecurity manager career path and some of the common interview questions that are asked.

Chapter 11, *Cybersecurity Sales Engineer*, is where you will learn about the cybersecurity sales engineer career path and some of the common interview questions that are asked.

Chapter 12, *Cybersecurity Product Manager*, is where you will learn about the cybersecurity product manager career path and some of the common interview questions that are asked.

Chapter 13, *Cybersecurity Project Manager*, is where you will learn about the Cybersecurity Project Manager career path and some of the common interview questions that are asked.

Chapter 14, *CISO*, is where you will learn about the **Chief Information Security Officer (CISO)** career path and some of the common interview questions that are asked.

Chapter 15, *Behavioral Interview Questions*, is where you will learn about some of the most common behavioral interview questions that are asked across cybersecurity career paths. This chapter is a must-read for anyone looking to be successful in their interview.

Chapter 16, *Final Thoughts*, is where we, the authors, share our final advice to help you succeed in both your job interview and in your cybersecurity career.

To get the most out of this book

To get the most out of this book, it's important to understand why you want to work in cybersecurity and to practice for your job interview. We suggest writing out the questions that you think the interviewer will ask you based on the job you are applying for, then do your best to answer those questions.

Doing this will help you during the job interview, ensuring that you are not stumbling around for answers to the interviewers' questions. This book is not intended to be read cover to cover, although you can do that. Instead, we suggest that you read *Chapter 1, Hacking Yourself, Chapter 15, Behavioral Interview Questions*, and *Chapter 16, Final Thoughts*, and then read only the chapters for the job roles that you are applying to. The information in all chapters will be beneficial to you, but by focusing on the job interview that is in front of you, you will be in a much better position to succeed. For example, if you are interviewing for a SOC Analyst job, the chapter on CISO interview questions will still be informative, but your main focus should be on the SOC Analyst interview questions, since that is the job interview you have next week.

After reading this book, it's critical that you actually apply the knowledge. People often say knowledge is power, but in reality, applied knowledge is the real superpower. The more you practice for your job interview in advance, the easier the job interview usually is.

Also, after you apply the information in this book and do well in your job interview, please share a post on social media and tag the authors because we care about your success and want to see your wins.

Download the color images

We also provide a PDF file that has color images of the screenshots/diagrams used in this book. You can download it here: `https://packt.link/gbp/9781835461297`.

Conventions used

There are a number of text conventions used throughout this book.

`CodeInText`: Indicates code words in text, database table names, folder names, filenames, file extensions, pathnames, dummy URLs, user input, and Twitter handles. For example: "You can use the `sleep` command, and if the web app sleeps for a period of time, it could indicate that it is vulnerable."

Bold: Indicates a new term, an important word, or words that you see on the screen. For instance, words in menus or dialog boxes appear in the text like this. For example: "I worked on a project integrating **CrowdStrike** with our **SIEM** system, using **RESTful** APIs."

 Warnings or important notes appear like this.

Tips and tricks appear like this.

Get in touch

Feedback from our readers is always welcome.

General feedback: Email feedback@packtpub.com and mention the book's title in the subject of your message. If you have questions about any aspect of this book, please email us at questions@packtpub.com.

Errata: Although we have taken every care to ensure the accuracy of our content, mistakes do happen. If you have found a mistake in this book, we would be grateful if you reported this to us. Please visit http://www.packtpub.com/submit-errata, click **Submit Errata**, and fill in the form.

Piracy: If you come across any illegal copies of our works in any form on the internet, we would be grateful if you would provide us with the location address or website name. Please contact us at copyright@packtpub.com with a link to the material.

If you are interested in becoming an author: If there is a topic that you have expertise in and you are interested in either writing or contributing to a book, please visit http://authors.packtpub.com.

Addendum — Salary Information Websites

The following websites for searching for cybersecurity jobs are provided for reference only. The authors have no affiliation with these websites, so we encourage you to use multiple sources of information and to run every website you find through a tool like Virus Total to quickly scan for potential threats. Some of the websites (e.g., LinkedIn Salary) can be used across multiple countries, so we have only listed these a single time. Again, this list does not contain every possible website out there for salary data, so please just use this information as a guide to help you get started.

Please also note that some of the links below may only work if you are located in that country.

United States

- Glassdoor: `https://www.glassdoor.com`
- Salary.com: `https://www.salary.com`
- PayScale: `https://www.payscale.com`
- Bureau of Labor Statistics (BLS): `https://www.bls.gov`
- Indeed Salary Search: `https://www.indeed.com/salaries`

India

- PayScale India: `https://www.payscale.com/research/IN`
- Naukri.com: `https://www.naukri.com`
- Glassdoor India: `https://www.glassdoor.co.in`
- Monster India: `https://www.monsterindia.com`

Vietnam

- Vietnam Works: `https://www.vietnamworks.com`
- Glassdoor Vietnam: `https://www.glassdoor.com.vn`
- CareerBuilder Vietnam: `https://www.careerbuilder.vn`
- JobStreet Vietnam: `https://www.jobstreet.vn`

United Kingdom (UK)

- PayScale UK: `https://www.payscale.com/research/UK`
- Glassdoor UK: `https://www.glassdoor.co.uk`
- Totaljobs: `https://www.totaljobs.com`
- Indeed UK Salary Search: `https://www.indeed.co.uk/salaries`
- Adzuna: `https://www.adzuna.co.uk`

France

- Glassdoor France: `https://www.glassdoor.fr`
- APEC: `https://www.apec.fr`
- Indeed France Salary Search: `https://www.indeed.fr/salaries`
- PayScale France: `https://www.payscale.com/research/FR`

South Africa

- PayScale South Africa: `https://www.payscale.com/research/ZA`
- Glassdoor South Africa: `https://www.glassdoor.co.za`
- CareerJunction: `https://www.careerjunction.co.za`
- Indeed South Africa Salary Search: `https://www.indeed.co.za/salaries`

Nigeria

- MySalaryScale Nigeria: `https://www.mysalaryscale.com/ng`
- Glassdoor Nigeria: `https://www.glassdoor.com.ng`
- Indeed Nigeria Salary Search: `https://www.indeed.com.ng/salaries`

Share your thoughts

Once you've read *Hack the Cybersecurity Interview*, we'd love to hear your thoughts! Scan the QR code below to go straight to the Amazon review page for this book and share your feedback.

`https://packt.link/r/1835461298`

Your review is important to us and the tech community and will help us make sure we're delivering excellent quality content.

Download a free PDF copy of this book

Thanks for purchasing this book!

Do you like to read on the go but are unable to carry your print books everywhere?

Is your eBook purchase not compatible with the device of your choice?

Don't worry, now with every Packt book you get a DRM-free PDF version of that book at no cost.

Read anywhere, any place, on any device. Search, copy, and paste code from your favorite technical books directly into your application.

The perks don't stop there, you can get exclusive access to discounts, newsletters, and great free content in your inbox daily.

Follow these simple steps to get the benefits:

1. Scan the QR code or visit the link below:

https://packt.link/free-ebook/9781835461297

2. Submit your proof of purchase.
3. That's it! We'll send your free PDF and other benefits to your email directly.

1

Hacking Yourself

In this chapter, you will learn how to use this book, especially if you only have a short period of time before your job interview. You will also learn about some of the most common job interview questions asked and recommendations on how to answer them. This chapter also covers a brief introduction to personal branding and a simple method to reduce stress before a job interview.

The following topics will be covered in this chapter:

- How to get the most out of this book
- General interview advice
- Common interview questions
- A definition of cybersecurity
- The **How, Analyze, Collect, and Know (HACK)** method
- Personal branding and soft skills
- Negotiation 101
- Managing stress

How to get the most out of this book

If you're reading this book and only have a short time before your job interview, let me share some wisdom on how to get the most out of it.

This book is not intended to be read from cover to cover, although it can be.

I would suggest reading through at least *Part 1* (*Hacking Yourself*) and *Part 4* (*Common Behavioral Interview Questions*), plus the chapter on the specific job role that you're interviewing for (assuming it's one listed in this book).

Many of the behavioral interview questions near the back of this book have been asked in interviews I've done, and some of the questions may be asked verbatim in your interview. This is why it's important to prepare for your interview, using both behavioral and technical interview questions, for the job that you want.

I want to stress that the technology mentioned in this book will likely change over the years, so the technical interview questions should not be taken as a *holy grail* guide but, rather, as a more general guide to the types of questions an employer may ask you during an interview. Also, as you read along with the topics in this book, based on your preferred job roles, you will come across some questions on various tools that are used in specific job roles. The questions you get about tools in an actual job interview may vary, based on the company and the specific role.

Being able to use a specific type of tool to solve a scenario is more important than trying to get experience in every possible tool on the market. For example, if the job description asks for **Splunk** experience and your experience is **QRadar**, you may not have time to become an expert in Splunk before the job interview. In this example, your skill in solving problems with a **Security Information and Event Nanagement (SIEM)** is what you should focus on, and then, in the job interview, speak about how you can learn new technology very quickly. This helps the interviewer see your skills and know that they can just send you through vendor training (in this example, Splunk) to build your skills for the specific tool they use.

We want to mention that this book could not possibly cover every possible interview question you might get, but we've made every effort to include some of the most common interview questions, after interviewing hundreds of cybersecurity professionals in these roles and going through hundreds of interviews ourselves over the years.

General interview advice

Your thoughts and words have power.

As strange as that statement might sound to you, I've found in life that statement is true.

I remember a few years ago, I needed to get my driver's license renewed and saw over 100 people waiting in front of me in the ticketing system line.

The first thought in my head was, *This is going to take all day*.

But what I said out loud to myself was, *This line is going to move quickly, and they will call my ticket number in less than 20 minutes. In fact, people ahead of me in the line are going to comment that this is the fastest they've ever seen the line move.*

Guess what happened.

Yes—you are correct.

The line moved quickly, and my ticket number was called in about 15 minutes. A few people ahead of me in the line also commented how they had never seen the line move that quickly.

Am I claiming some superhero power, or am I planning to dive into a deep metaphysical discussion here?

Not at all—and I would much rather have the ability to freeze someone with ice as a superpower, anyway.

However, this is an example of the power of your words. You can search online and across social media platforms to see thousands of other examples of this.

 There are also numerous books on the subject. Some good ones that I've read are *What to Say When You Talk to Yourself* by Shad Helmstetter, *The Power of Awareness* by Neville Goddard, and *You Are the Way* by Fabio Mantegna and Elmer O. Locker Jr.

It's important to speak the right way before any job interview. Instead of saying things such as *I'm dumb* or *They will hate me*, say things such as *This is going to be a great interview. Everyone is going to be friendly to me. In fact, it will feel more like a conversation with old friends than a job interview*. It also helps if you write out exactly how the interview is going to go as if you had already experienced your perfect job interview. This is the power of visualization.

Now, does this guarantee you will ace the interview and get the job offer? No, of course not. And sometimes, you don't get certain jobs because something much better for you is right around the corner. I have even experienced this in the past.

Common interview questions

Now, let's look at some common interview questions you may be asked during the initial phone screen interview, the hiring manager interview, and/or any team interview:

Tell me about yourself

This question is not a place for you to share your childhood memories. It's also not a place for you to mention generic stuff like *I'm a lifelong learner* because everyone in a cybersecurity career is a lifelong learner, due to technology and threats constantly evolving.

Instead, what the interviewer wants to hear is a brief summary of your career with a focus on your *impact* on past employers. Think of this as your 30-second elevator pitch. Be as specific as possible in your answer.

Here's a formula to help you craft a good answer to this question:

- A one-sentence introduction to who you are professionally
- Two to four metrics that make you stand out based on the job role
- One or two sentences about why you want this job

Here's an example of the formula being used:

I have been in healthcare cybersecurity for 3 years and currently work as a Cybersecurity Analyst for ABC Hospital, where I built automation that reduced our support tickets by 11.2%. This opportunity caught my eye because the company's mission to bring better healthcare to rural communities is a passion of mine, and I can make an immediate impact on the cybersecurity team. Would you like to hear more about anything I've mentioned so far?

The reason I suggest answering using this formula is because it keeps your answer concise. Asking the interviewer if they want to learn more about what you've mentioned both respects their time and allows them to dig deeper into areas of your response that are the most important to them. The problem that many job candidates have if they don't use a formula like this is that they ramble on for 5 to 10 minutes. This means the interviewer is now in a rush to finish their list of questions, which leaves little time for you as the job candidate to get your questions answered at the end of the interview.

Where do you see yourself in 5 years?

With this question, the interviewer is trying to determine if you have a plan for your future with the organization and in your career.

Most people will answer this question with something about taking the interviewer's job or becoming a **Chief Information Security Officer** (**CISO**) in 5 years even if they have no cybersecurity experience, but the best way to answer is to mention that you want to master the current job and become the go-to person in your domain, and then speak about how the job aligns with your long-term goals.

Here's an example:

I want to take the next year to fully learn my role and the company's needs, and then, in the following years, establish myself as a thought leader in incident response. This job as a SOC Analyst would help me build the foundational skills for incident response.

What is your greatest strength?

Under the section titled *The HACK method* later in this chapter, we'll talk about analyzing yourself, which is what you need to do to answer this question. In the meantime, for this question, talk about a specific strength and a specific example of when you used your strength.

Here's an example:

I would say my greatest strength is the ability to communicate effectively with different stakeholders. An example of this is when I reported on an incident to senior leadership and communicated the effectiveness of our existing security controls, while recommending areas we could improve. This helped my team get additional funding to implement some of the suggested security controls.

What is your greatest weakness?

With this question, you want to mention your greatness weakness, what you are doing to resolve it, and the status of your resolution.

Here's an example:

I would say my greatest weakness is being nervous with public speaking, so I started speaking at a Toastmasters group, and I am also taking an online course on public speaking best practices. So far, I have given three talks at my Toastmasters group, and I'm about halfway through the public speaking course. I feel much more comfortable now in front of large crowds.

Why are you leaving your current job?

I usually would answer this one with something about growing your career into the new role.

Here's an example:

While I've enjoyed my time at Acme Inc. I realized it was time to move to the next level in my career as a pentester. (Depending on the role, you can adjust your response. For example, if you are moving from an individual contributor role into a manager job, mention that you are looking for more responsibility in a leadership role.)

What are your salary expectations?

These days, many job postings will list the base salary range. If the range is not listed in the job description, you might be asked on the job application or during the job interview what your salary expectations are. This question trips a lot of people up. Instead of giving them a salary range, which puts you at a disadvantage in the negotiation stage, I suggest using the following example:

I'm open to discussing compensation at the right time. I'm looking for the best overall fit and package. Are you opposed to sharing your budgeted range for this role with me?

If they refuse to share their budgeted range with you, which is rare, then it might be time for you to explore working for another company.

At some point in the interview process, typically near the end of the interview, you will be asked if you have any additional questions. During this time, some job candidates say they don't have any questions because they think that will help them look smarter to the interviewer. The reality is that the interviewer expects you to ask questions to ensure the company is a good fit for you.

Job candidates will also ask about the company culture at this stage of the interview as well, but I suggest you ask the more specific questions below first because it will help you identify exactly what the interviewer is looking for in the candidate they ultimately hire for the job.

Here are some interview questions I think you should ask them:

What seems to be missing from the other candidates you've interviewed so far?

This question helps you identify exactly what the interviewer is looking for and the gaps other candidates have. Knowing this information helps you speak about how your experience fills those gaps and why you are the best candidate for the job. This question should be the first one you ask when they give you time to ask them questions, and it can be asked during every stage of the interview because you want to know what each person is looking for in the perfect candidate. Just asking this question will often help you move to the next round of interviews because most people do not ask intelligent questions like this one. They typically ask about company culture, but you should do your research and know about the culture before the job interview.

Which key performance indicators (KPIs) would you have for me in this position over the first 30, 60, and 90 days?

This question does a few things that benefit you. First, it gets the interviewer thinking of you in the position and not other candidates.

Second, it lets you know what is expected of you in the position over the next 90 days. If the interviewer (assuming it's the hiring manager) has not thought about any KPIs for the position, it might indicate they are overwhelmed with work, and it might not be a good company for you. See the final chapter of this book for a suggested slide presentation, where you can come up with some of the things you plan to do in the first 30, 60, and 90 days of the job. By doing the slide presentation, you can help hiring managers figure out what to do with you when you start the job. Most other candidates don't take these extra steps, and this is a good way to impress the hiring manager.

What kind of person succeeds at this company?

This helps you understand how the company defines success. If the answer is someone who works 100-hour weeks, then you should probably run out of that interview as fast as you can.

What do you enjoy most about working at this company?

If the interviewer is happy in their role, then they might share a few things they love. On the other hand, by asking this question, you might be able to save yourself some headaches from working in a toxic environment. It's amazing what some interviewers will share with you if you ask the right questions (social engineering at its finest).

What does it mean to be a culture fit at <company name>?

This question is a simple way for you to find out from an insider exactly what they are looking for as the "right fit" for the job. However, before asking this question, you should always ask the interviewer what seems to be missing from other candidates.

Some additional tips for your interview

Here are some additional aspects to keep in mind during your interview:

- **Make eye contact**: I would say you want to make eye contact most of the time when you're listening to the interviewer and when you're answering questions. Little or no eye contact can make people suspicious and feel you're not trustworthy.
- **Smile more**: Don't be creepy with this one, though. If someone is constantly smiling in the interview, I immediately feel it's not genuine. Smile when appropriate, and if you're introverted like me, then try to remember to smile at least three to four times during the interview, especially when you first meet the interviewer.

- **Appearance**: As much as that person you follow on social media might want you to believe that appearance doesn't matter, it does. Be sure that you're well-groomed (and showered, please) and dress appropriately. I do recommend a suit (men and women) if you have one, but in most interviews, business casual is fine. I would suggest asking the person who set up the interview what the dress code is. You want to maintain the same dress code for virtual interviews because you don't know when you may find yourself standing up during the interview and being caught just wearing your shorts—or worse.

- **Research the company**: Do your homework on the company—its mission, current/future project initiatives, financials, and so on. I'm always amazed at how many people show up to an interview without having done any **open-source intelligence** (**OSINT**) on the company. I had one person walk past the sign with the business name on it and then ask me what company it was in the job interview. Guess what? They didn't get hired. Avoid being lazy like that person and do your research.

- **Don't bad-mouth a past employer or team**: Yes, some companies (and some people) are not the best for you, but no one wants a negative person on their team. I remember a person I worked with many years ago who was negative about everything, and several productive people left the team because they were tired of hearing endless complaints.

  Remember, it only takes one bad apple on a team to change team dynamics and reduce the team's productivity.

- **Don't be emotional**: Remember, this is business, so don't get emotional when talking about past companies, and so on. The interviewer is not your therapist. For example, let's say a past boss mistreated you. Instead of showing anger or crying during the job interview about the past situation, calmly speak about the situation you experienced in the past, how you recognized that place was not the right fit for you, and what you like about this new company.

- **Be concise in your answers**: For most people, this means you need to practice your answers to common interview questions and figure out how you can say less to get the same point across. We have a chapter on behavioral interview questions. I suggest using **Problem, Action, Result** (**PAR**) to answer these types of questions—what was the problem you were solving for, who was involved, what did they do, and what was the end result of the situation?

Some job candidates may find it challenging to be concise in their responses. If this is you, here are some things that I used to do during a job interview to reduce my rambling:

- *To cut a long story short* is a phrase I use if I think I'm rambling so that I can wrap up whatever I am saying. You must practice this and be conscious of the fact you are rambling for this one to be effective.

- **Keep your answer short**. I used to answer interviewer questions with just a few words and then ask them if they wanted to know anything more about what I had mentioned.

- The words *but* and *because* are also helpful in your interview.

The word *but* can help explain something you lack and why you are still the right candidate for the job. For example, *I don't have the five years of required incident response experience,* **but** *I do have several years of experience in Splunk and managing incidents.*

In human psychology, people often only remember everything after the word *but*, and everything after it is used to justify why they should do something. In this example, you give them a reason to consider you for the job because you do have experience in Splunk, which was also required in the job description.

The word *because* is helpful in compensation negotiations. For example, let's say you want to make $100,000 USD and they offer you $80,000 USD, but you know the salary range for the job is up to $120,000 USD.

After they make you the offer, you could say something like *I appreciate the offer, but it seems like we should be looking at around $100,000 USD* **because** *I have several years of experience with Splunk and in incident response.*

A definition of Cybersecurity

Now that you have some basic tips for your interview, let's talk about this whole cybersecurity thing in case you're new to the field.

If you ask 100 people the definition of cybersecurity, you'll get 100 different answers.

The **National Institute of Standards and Technology (NIST)** defines cybersecurity as follows:

Prevention of damage to, protection of, and restoration of computers, electronic communications systems, electronic communications services, wire communication, and electronic communication, including information contained therein, to ensure its availability, integrity, authentication, confidentiality, and nonrepudiation (https://csrc.nist.gov/glossary/term/cybersecurity).

We cover several cybersecurity career paths in this book. Here is an overview of how each career path ties into overall cybersecurity for organizations.

Cybersecurity Engineer

A Cybersecurity Engineer designs, implements, and maintains security systems and controls to help protect organizations from various cybersecurity threats.

SOC Analyst

A SOC Analyst monitors network and system activities to detect and respond to security incidents.

Penetration Tester

Penetration Testers (sometimes called ethical hackers) identify vulnerabilities in systems and networks and attempt to exploit them before they can be exploited by malicious actors. This helps organizations take a more proactive approach to their cybersecurity.

Digital Forensic Analyst

Digital Forensic Analysts secure, collect, and analyze data from digital devices following cyber incidents. This role is important in the incident response phase. It helps an organization understand what happened during an attack so that the incident response team can mitigate future attacks.

Cryptographer/Cryptanalyst

Cryptographers and Cryptanalysts develop encryption algorithms, which can help organizations protect their data if it is lost or stolen.

GRC/Privacy Analyst

GRC Analysts, Privacy Analysts, and other GRC roles help ensure that organizational practices and data handling meet regulatory compliance requirements and best practices. These roles help in preventing incidents and ensuring compliance by establishing policies and procedures that protect data and reduce risk.

Security Auditor

Security Auditors assess the security posture of an organization by reviewing systems and processes to identify vulnerabilities. This role focuses on prevention and compliance by ensuring that existing security measures are adequate and recommending improvements.

Malware Analyst

A Malware Analyst analyzes malware to identify its purpose and **indicators of compromise** (**IOCs**) from the malware, which can help the organization protect against future attacks.

Cybersecurity Manager

Cybersecurity Managers oversee the implementation of security policies and procedures across an organization. This role is involved in all aspects of cybersecurity, including prevention, detection, and response, by managing the overall cybersecurity team and implementation strategies.

Cybersecurity Sales Engineer

Cybersecurity Sales Engineers work with clients to design and sell security solutions tailored to their organization's needs. This role contributes to cybersecurity by ensuring that clients have the necessary tools to help protect themselves against emerging threats.

Cybersecurity Product Manager

Cybersecurity Product Managers lead the development of cybersecurity products through the entire product lifecycle. This role helps client organizations by building security products that meet client needs.

Cybersecurity Project Manager

As the name implies, Cybersecurity Project Managers are responsible for ensuring that projects are executed on time and within budget and achieve their intended security outcomes.

Chief Information Security Officer (CISO)

The CISO leads strategic planning and governance of cybersecurity across an organization. The CISO is key in all aspects of cybersecurity, setting the vision and strategy to protect the organization from cyber threats and ensuring compliance with security regulations.

The HACK method

I could feel beads of sweat forming on my forehead as I stared at my computer screen in the darkness of the night. My stomach churned as I watched the timer count down and I thought, *Will I make it in time?*

Was I doing some top-secret hack against an alien spaceship to save the world?

No—I had simply procrastinated in writing some papers for my classes, and I now had less than an hour to write three lengthy papers.

Besides, I didn't have my hoodie and gloves on, which we all know is a requirement of any successful hacker (if you don't understand the joke here, just google it).

With a few minutes left, I submitted all three papers and received an excellent grade on all of them.

What was my secret? Did I hack into the professor's computer to change my grades? In hindsight, that might have been a good option, but instead, I had learned a long time ago how to hack myself.

In a similar fashion, you can learn how to **hack** yourself for job interviews using the simple HACK method.

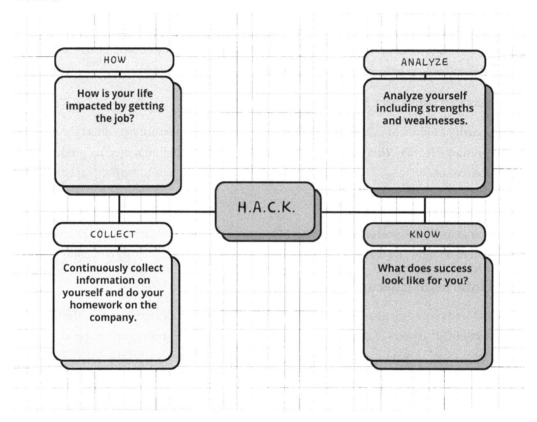

Figure 1.1: The HACK method

Have a look at the following definitions of the HACK method.

How

When applying for jobs, think about how your life is impacted by getting this job. Many people will just focus on the money here, but they should also ask themselves how this job fits into things such as their long-term plans. What sacrifices do you have to make (missing birthday parties, a long commute to the office, and so on) for this job?

Here are a few things I used to always think about:

- *How long is the commute?* I once had to commute 4+ hours each way for a job, so I'd spend much of the week just sleeping in my car near the job site.

- *How much earning ability do I have?* Is this just a base salary, or is there an option to earn more with stock options, sales commission, and so on?

- *How much time will I really spend on this job?* Because most of us are paid for 40 hours a week in the US but actually work 60+ hours each week.

- *How does this job benefit my 1-, 5-, and 10-year goals?*

Analyze

The next part of HACK is analyzing yourself. There are a number of self-assessments out there on the internet, all sorts of personality and aptitude assessments, and so on.

Those are fine to take, but the low-cost route is to get a piece of paper (or your phone notes) and write down what you think is important, how you work through problems/projects, how good you are at time management, and so on.

This is important, so be honest with yourself.

As an example, I do the self-analysis monthly and I know that I am willing to go without eating, sleeping, and entertainment so that I can finish a project. I have no hesitation in sacrificing to complete the *mission*, which is how I was able to write those 20+-page papers in a short period of time. I also know that I can complete projects quickly, so I sometimes procrastinate until the deadline.

Collect

It's important to collect information about yourself continuously. The good news is, you'll likely secure that information better than the large companies out there.

It's also important for you to collect information about the job you are applying for, the company itself (as mentioned earlier in this book), and what your long-term goals are.

Know

You need to know what success looks like for you. For example, one cybersecurity professional I know makes a relatively low salary but can finish at 5 p.m. every day to have dinner with his family. He's happy with his job, and that is what defines success for him.

Another cybersecurity professional I know is single and works at a major tech company, working 80+ hour weeks but making close to a million in total compensation. This is the definition of success for her.

It's crucial to know what success looks like for you and not what people on social media tell you success is.

If you built a life that you never had to take a vacation from, what would that life look like?

Personal branding and soft skills

Many of the hiring managers I have spoken with over the years have mentioned that soft skills are a key part of the ideal candidate. In this section, we will discuss how personal branding and soft skills can help you achieve your goals. Your personal brand can help you develop more confidence, provide better job security, and increase your earnings over your career. When I developed my personal brand years ago, I went from living paycheck to paycheck to having financial freedom.

Why does personal branding matter for job interviews? Building your personal brand does a few things for you:

- Helps you get more interviews because you are looked at as a thought leader in your specialty
- Helps you build confidence, which in turn helps you have more confidence in job interviews
- Helps you when it's time to negotiate your salary and other compensation

Personal branding

"Two all-beef patties, special sauce, lettuce, cheese, pickles, onions on a sesame seed bun" was a commercial jingle from McDonald's in the 1970s. I first heard it years later and I still remember it now.

That's the power of proper branding.

You are a brand, and you have value. Your personal brand brings value to any organization and helps them make more money.

Let me ask you a question. If you work hard for your brand and a company makes more money because of your brand, wouldn't it make sense for you to earn more money as well? Hopefully, your answer is yes.

There are many books on personal branding and many ways to build your personal brand. I typically would tell you to focus on **LinkedIn** and show what you know through videos/screen recordings, posts, or articles/whitepapers.

Your personal brand can help you get jobs.

Adopt the *no spray and pray* résumé and job application approach—then, you'll have no *we have a few more candidates to interview* type of responses and no real pushback on the salary you want. All of this is made possible by your personal brand. For those not familiar with the *spray and pray* job application approach, it just means that you apply to hundreds of jobs and hope that you hear something back. By building your personal brand, you can get companies reaching out to you instead of you applying to hundreds of jobs on job board websites like Indeed.

Soft skills

You might hear of many companies that advertise that they need people with soft skills. Some of the key soft skills I think someone needs to have are the ability to communicate effectively across different stakeholders, the ability to work in a team, an **emotional quotient** (**EQ**), which is also known as emotional intelligence, and customer service skills.

Negotiation 101

Everyone is in sales is a statement from one of my mentors many years ago.

Would you be opposed to me teaching you a trick I have used over the years to win at job interviews and—especially—negotiations?

Most people would answer *no* to that question because it's easier for most people to answer naturally with a *no* to a question than a *yes*. Also, people want to know what the interview trick I have is, so they would answer *no*, meaning that they are not opposed to me sharing that trick with them.

I won't deep dive into sales techniques or human psychology in this book, but a good sales book is *The Sandler Rules* by David Mattson, and a guy named Josh Braun also has some good training. In addition, the book *Pre-Suasion* by Robert Cialdini can be helpful for understanding sales psychology.

Here are a few questions I have used over the years in interviews:

Would you be opposed to...?

Here's an example:

Would you be opposed to me asking about career advancement for this job role?

Would it make sense...?

Here's an example:

Would it make sense for us to discuss salary after we've seen that this is a good match?

Can you offer your advice on…?

Here's an example:

Can you offer your advice on how the team manages projects?

It seems…?

(Note: shut up after you use this one and let the interviewer respond.)

Here's an example:

It seems like you need a minimum salary expectation to move me forward in the application process.

Managing stress

Years ago, I read a book called *How to Stop Worrying and Start Living* by Dale Carnegie. I'm not going to share everything in that book (you should buy a copy), but one key exercise had you reflecting on the worst possible scenario that could happen and then asking yourself if you were OK with that happening. If your answer was no, then you had to think through what action you could take to improve the situation.

Here's an example:

- **Scenario**: Your boss wants a project done by Monday, even though the real deadline is 3 weeks away.
- **Worst-case scenario**: You don't do the project on your days off, and the boss fires you because of it.

Are you OK with this?

Yes, because you already have money saved up to cover expenses while you look for another job. This gives you some freedom, and typically, this scenario wouldn't lead to termination, since the project is not due for weeks.

No—you really need this job to pay your bills. In this situation, you probably have to suck it up this time and get the project done. However, I would suggest you then focus your spare time on building additional income streams, saving more money, and/or finding a new job. Otherwise, the cycle will just repeat itself.

Some information online suggests that reading a book can help reduce your stress. I've personally found that reading can help take your mind away from stressful situations.

Another good book on the subject is *Chaos Loves You: So Let's Love It Back* by Jothi Dugar (a cyber-security executive).

I also practice simple meditation and breathing exercises.

Going back to what I mentioned earlier in the book as well, how you talk to yourself is crucial to your success and in removing stress from your life.

I'm not going to lie to you—cybersecurity careers can be extremely stressful. It's important for you to recognize if you're stressed out and identify safe ways to cope. Remember, we as a community are here to support you.

Now that you have a good idea of some common interview questions you might be asked, the questions you should ask, the HACK method, and a few books that I suggest you buy about stress management, let's move into job-specific interview questions in the upcoming chapters.

Summary

In this chapter, you learned about common interview questions asked in many job interviews. Studying these questions and writing down your answers to them in advance of a job interview can help the interview be less stressful. Speaking of stress, you also learned a simple method to analyze a situation to help lower your stress level about the situation. You also learned how to *hack* yourself for job interviews. Understanding yourself and your situation (situational awareness) is critical to your success in job interviews and life. In addition, you learned the importance of building your personal brand, some of the key soft skills you need, basic negotiation skills, and simple ways to manage stress.

In the next chapter, you will learn about Cybersecurity Engineer careers and some common knowledge questions you might be asked in a job interview.

Join us on Discord!

Read this book alongside other users. Ask questions, provide solutions to other readers, and much more.

Scan the QR code or visit the link to join the community.

`https://packt.link/SecNet`

2

Cybersecurity Engineer

In this chapter, you will learn what a **Cybersecurity Engineer** is and the average salary range for this career in the United States. You will also learn about the career progression options and learn common interview questions for the role. Each potential job title has interview questions listed that you might see in your job interview; however, the questions you get in one might be different than those listed.

The following topics will be covered in this chapter:

- What is a Cybersecurity Engineer?
- How much can you make in this career?
- What other careers can you pursue?
- Common interview questions for a Cybersecurity Engineer career

What is a Cybersecurity Engineer?

Cybersecurity Engineers are responsible for building secure infrastructure that helps protect organizations from security threats. Alongside that, they are also involved in security testing activities, like **vulnerability scanning** and building and implementing **security policies** and procedures, as well as in **incident response** and ensuring compliance.

The specific job tasks you do as a **Cybersecurity Engineer** might vary based on the organization because each organization has different security goals, business objectives, and technology stacks. Other factors like industry best practices and compliance requirements can impact how an organization approaches security, which can impact your daily tasks as a cybersecurity engineer.

These factors can lead to Cybersecurity Engineers performing tasks that could range from network security to compliance and being involved in audits. It is important to look at the job description carefully to see what skills the company is asking you to have and what some of your key tasks will be.

For example, your job title might be Product Security Engineer, and in this role, you would work closely with the software development team to build more secure software. One of your friends might have the job title, Cybersecurity Engineer and be focused on building secure network architecture for their organization.

There are several cybersecurity roles that might include responsibilities from cybersecurity engineering. These other possible job titles include the following:

- **Security Engineer**: A security engineer designs, implements, and manages security solutions to help protect an organization's digital assets and infrastructure from threats.
- **Product Security Engineer**: Product security engineers develop and integrate security measures specifically for software products, ensuring that they are safe from cyber threats throughout their lifecycle.
- **Cybersecurity Architect**: A cybersecurity architect designs comprehensive cybersecurity architecture that aligns with business needs, while effectively mitigating security risks.
- **Security Automation Engineer**: A security automation engineer develops automated tools and systems to help security operations teams perform their tasks efficiently. This helps the teams focus on the most important alerts.

It is important to keep in mind that the organization you wish to work at might have a different job title and job responsibilities than the ones listed in this book. Cybersecurity Engineers work closely with various teams across an organization, ensuring that security measures are integrated into all aspects of business operations and technology infrastructure. This collaboration is important because security is not just a technical issue but also a business one, impacting everything from compliance and risk management to operational efficiency and reputation.

How much can you make in this career?

Cybersecurity Engineer salaries can vary significantly by location, company, and other factors. In the United States, you can expect to make between $72,000 and $146,000+ for a Cybersecurity Engineer position. The alternate job titles listed above can have a wide range of compensation available. It's important to search for the specific job title and company you want to work at to see the estimated compensation.

What other careers can you pursue?

A career as a Cybersecurity Engineer builds a solid skill set and can help you prepare for many cybersecurity careers. Some examples are:

- Malware Reverse Engineer
- Penetration Tester
- Cloud Security Engineer
- Cybersecurity Manager

Common interview questions for a Cybersecurity Engineer career

In the following section, you will learn about the common interview questions for a Cybersecurity Engineer. I have also included a few behavioral interview questions and explained what the interviewer is looking for by asking the question. This book has an entire chapter dedicated to behavioral interview questions, so please be sure to read through that chapter before your next job interview.

Remember that *clear and concise answers make the interview nice.*

General Cybersecurity Engineer knowledge questions

In this section, you will see some general questions that might be asked in a Cybersecurity Engineer job interview. The following questions are separated by job title and reflect questions you might be asked, based on real-life job postings.

Security Engineer Interview Questions

Can you describe a cybersecurity incident you have resolved in the past and explain the steps you took to mitigate risk?

Example answer:

In a previous role, I encountered a massive, **Distributed Denial-of-Service (DDoS)** attack. I identified the attack vectors through real-time monitoring and log analysis, implemented rate limiting, and deployed additional firewall rules to mitigate the attack. Post-incident, I led a review that resulted in an enhanced DDoS mitigation strategy, including better traffic analysis and response plans.

How do you manage and secure Microsoft environments, specifically with MS Defender products across different platforms, such as O365, cloud, and identity management?

Example answer:

I have managed Microsoft environments by leveraging MS Defender across various platforms. For example, in O365, I ensured the configuration of Defender for Office 365 against phishing and malware. For cloud environments, I implemented Defender for Cloud to secure Azure services, and integrated Defender for Identity to protect against identity-based threats.

Explain how you have utilized the National Institute of Standards and Technology (NIST) framework in a previous role to improve a security posture. Can you provide a specific example of a policy or procedure you developed based on NIST guidelines?

Example answer:

At my previous job, I integrated the NIST Cybersecurity Framework by aligning our security policies with its core functions: Identify, Protect, Detect, Respond, and Recover. I developed an incident response strategy that reduced our mean time to detect and respond to incidents by 30%, significantly enhancing our resilience to cyber threats.

Can you give an example of a security policy you wrote?

Example answer:

I led the build of a new data encryption policy that required the use of **AES-256** encryption for data at rest. It also required the organization to use **TLS 1.2** or higher for data in transit. I also helped ensure we remained compliant by holding quarterly training sessions with the team and used continuous monitoring solutions to ensure everyone was following the policy.

How do you administer and monitor security profiles and policies?

Example answer:

I review access to ensure only the minimum amount of access needed to perform a function or task is used. I also use tools like **Security Information and Event Management (SIEM)** to monitor and analyze security logs and aggregate this data in a centralized dashboard. In my last role, I led an investigation team that investigated policy and access violations.

Product Security Engineer Interview Questions

Since the job responsibilities for a Product Security Engineer can vary so greatly across organizations and industries, the question examples below focus on a Product Security Engineer role at a healthcare organization.

Can you describe a scenario from your experience where you identified a critical vulnerability in a medical device? How did you assess the risk, at a high level, and what steps did you take to mitigate it?

Example answer:

In my previous role, I identified a **buffer overflow** vulnerability in a **defibrillator**'s software. I conducted a risk assessment using a threat modeling approach, determining that the vulnerability could allow unauthorized access to device settings, which meant an attacker could turn off the defibrillator function, thereby risking patient safety. I worked with our software development team to redesign the input validation process and implement secure coding best practices. We then conducted testing of the changes to ensure that the vulnerability had been resolved.

Tell me about a time when you influenced the architecture and design of a product to enhance its security. What were the security considerations you ensured were incorporated?

Example answer:

In a project designing a new insulin pump, I led the security architecture discussions, ensuring that all security considerations were integrated. I advocated for and implemented secure communication protocols and encryption for data at rest and in transit, helping us ensure compliance with both safety and privacy regulations.

How do you ensure your design documentation meets the industry standards for medical device software, such as IEC 62304 (`https://www.iso.org/standard/38421.html`)? Can you describe the process you follow?

Example answer:

For compliance with **IEC 62304**, I maintain thorough documentation throughout the software development process. This includes detailed design specifications, risk analysis reports, and validation and verification plans.

Regular audits and reviews by a separate team ensure that all documents meet the stringent standards required for medical device software, as outlined in **IEC 62304**.

Give an example of a security solution you implemented in a medical device. What challenges did you face, and how did you overcome them?

Example answer:

I led a team that implemented a **multi-factor authentication** (**MFA**) solution in a wearable health device, which was challenging due to device limitations and user interaction constraints. My team overcame these challenges by using lightweight cryptographic protocols and optimizing the authentication process to balance security with user convenience.

Cybersecurity Architect Interview Questions

Can you describe your experience with Security Operations Center (SOC) technologies, particularly SIEM and SOC automation, and how did you implement these technologies in past projects to reduce incident response times?

Example answer:

In my previous role, I implemented an SIEM solution that integrated with existing SOC automation tools to streamline our incident response. This included setting up correlation rules that automatically detected anomalies and triggered security workflows, reducing our response times by 17%.

Given your cross-domain knowledge and experience, can you discuss how you integrated endpoint security and identity and access management (IAM) solutions in a previous role to improve an organization's overall security posture?

Example answer:

I integrated endpoint security with IAM by deploying unified endpoint management that enforced device compliance, before granting access to corporate resources. This approach reduced the attack surface and improved the security posture by ensuring consistent security policies across all devices.

How do you approach building security architectures that span multiple cloud platforms? What challenges did you face in the past, and how did you address them?

Example answer:

I have designed security architectures across AWS, Azure, and GCP by utilizing each platform's native security tools and ensuring that all configurations adhere to best practices. My approach often involves using a centralized security management tool to ensure visibility and control over all platforms.

Describe your experience designing security for hybrid environments that include on-premises, co-located, and cloud-hosted architectures. What specific strategies did you employ to manage security across these varied environments?

Example answer:

For a hybrid environment, I developed a security strategy that included unified threat management, providing seamless security across on-premises and cloud components. Key tactics included consistent encryption policies and the use of **cloud access security brokers** (**CASBs**) to monitor and control data movement.

Security Automation Engineer Interview Questions

Can you describe your experience with Security Orchestration, Automation, and Response (SOAR) platforms? Specifically, how have you developed and deployed playbooks in your previous roles?

Example answer:

In my previous role, I utilized Splunk Phantom extensively to create automated playbooks. I developed a playbook for incident response that automated the initial triage of alerts, gathered additional context from various sources, and executed predefined mitigation steps. This reduced our average response time from hours to minutes and significantly decreased manual efforts.

Tell me about a script you wrote to automate a security process? What was the challenge, and what impact did your script have?

Example answer:

I created a Python script that automated the process of log collection and parsing across multiple systems, which are part of our security operations. The script consolidated logs in a central repository where further analysis could be conducted. This automation saved time and improved our log management process's consistency and reliability.

Describe a time when you integrated multiple cybersecurity vendor tools using APIs.

Example answer:

I worked on a project integrating **CrowdStrike** with our SIEM solution, and we used **RESTful** APIs. The main challenge was ensuring that the data from CrowdStrike's EDR was ingested in a format that the SIEM could manage. To solve this challenge, I developed a middleware layer that put the data into a format that could be read by the SIEM.

How have you used cyber threat intelligence in the context of security automation to mitigate threats?

Example answer:

I used SOAR with our threat intelligence feeds to give more context to the information we were seeing. By pulling contextual information automatically, my team could prioritize incidents more accurately (reduce false positives and false negatives) and respond faster to active incidents. One example of how this proved to be helpful is during a ransomware attack, where having this additional data helped the team respond faster and isolate the affected systems.

Give me an example of how you have automated security across a public cloud environment.

Example answer:

In the AWS cloud, I automated security group audits and remediations. I did this by using Lambda functions triggered by scheduled events, which ensured the system would verify compliance with our security policies and adjust security groups automatically to close any unauthorized access.

Walk me through your approach to developing automated workflows for security operations. How do you ensure these workflows are effective and efficient?

Example answer:

To develop effective automated workflows, I use a combination of process mapping and pilot testing. Each workflow is initially mapped out, with existing manual processes considered, and then tested in a controlled environment. Adjustments are made based on performance metrics and feedback from stakeholders, and everything is tested again before full deployment in production.

Tell me about a time you had to change legacy systems or processes in an organization. How did you approach stakeholder management and ensure the transition was smooth?

Example answer:

To transition from legacy processes, I focus on comprehensive stakeholder engagement and clear communication. For example, when automating data extraction processes, I conducted workshops with the IT team to understand their concerns and requirements, ensuring the new system addressed these specifications.

In addition to the technical questions that may be asked for a specific job role, you might be asked how you stay up-to-date with trends and emerging threats in cybersecurity:

How do you stay current on cybersecurity trends?

The answer to this question depends on which sources you use for cybersecurity news and trends. The interviewer is just looking to see if you stay up to date on things that are happening, as competent security professionals must remain current on the latest threats that could impact their organization.

Some sources of information include new websites, social media (i.e. LinkedIn or X), blogs, podcasts, white papers, your peers, and newsletters.

The goal here is not for you to try and consume every possible piece of cybersecurity-related content out there. The goal is to just ensure that you have some method to stay current on emerging threats. For example, you might find that the interviewer and you have a shared favorite podcast. This shared interest can help you overcome the similar-to-me bias that some interviewers have.

Summary

In this chapter, you learned about the Cybersecurity Engineer career and the average salary range in the United States. There are several job titles you might see online that have cybersecurity engineering responsibilities, including Security Engineer, Product Security Engineer, Cybersecurity Architect, and Security Automation Engineer. You also learned how cybersecurity engineering roles can be a stepping stone to other cybersecurity careers, like malware reverse engineer, and you learned common interview questions asked for Cybersecurity Engineer roles.

In the next chapter, you will learn about a career as a SOC Analyst, including common knowledge-based interview questions you might be asked.

Join us on Discord!

Read this book alongside other users. Ask questions, provide solutions to other readers, and much more.

Scan the QR code or visit the link to join the community.

`https://packt.link/SecNet`

3

SOC Analyst

In this chapter, you will learn what a **Security Operations Center (SOC)** analyst is and the average salary range for this career in the United States. You will also learn about the career progression options and common interview questions for the role.

The following topics will be covered in this chapter:

- What is a SOC Analyst?
- How much can you make in this career?
- What other careers can you pursue
- Common interview questions for a SOC Analyst career

What is a SOC Analyst?

SOC analysts work as members of a managed security services team. There are typically three tiers of SOC analysts, and job-specific duties may vary based on the organization you work for:

- **SOC level 1 (tier 1) analysts** typically monitor security tools, such as **Endpoint Detection and Response (EDR)** and **Security Information and Event Management (SIEM)** tools, to identify potential anomalous activity on networks and systems. If anomalous activity is detected, they then escalate it to level 2 analysts.

- **SOC level 2 (tier 2) analysts** investigate anomalous behavior. In some instances, they may perform **incident response (IR)** duties and initial malware analysis. You might build IR playbooks and perform scripting to automate routine tasks. You might also see level 2 skills being requested for incident responder job postings. Your tier 2 SOC Analyst might also set up the access for jump boxes and do light forensic investigation work.

- **SOC level 3 (tier 3) analysts** perform IR and also typically perform threat hunting and threat profiling. They may also do some work in reverse engineering malware and digital forensics, depending on their organization. You might see these job openings listed as incident responders or threat analysts/hunters. One thing to keep in mind if you are transitioning from another career to cybersecurity is you can often find non-traditional jobs at a cybersecurity product company, using this as the starting point for your career. As an example, if you are transitioning from selling used cars, you could get a job with the sales team at a security company such as Splunk. The company will then train you on all of their cybersecurity product and service offerings for free, and then in 6 to 12 months, you will have a better chance of getting a cybersecurity job because you will then have experience at Splunk as well as experience with the company's different product offerings, so you will have in-demand skills. Many people focus on getting jobs as a SOC Analyst or Penetration Tester because that's what their guidance counselor tells them to do, but it is often a better idea to look at non-traditional jobs to get your start in a cybersecurity career because others will not apply for those jobs. If you look at the Splunk company website, you will see hundreds of open non-traditional jobs (As of June, 2024) that can be leveraged to get your start in a rewarding cybersecurity career.

How much can you make in this career?

In the United States, you can expect to make between $58,000 and $88,300 for an entry-level SOC level 1 role. Some SOC analysts working in the U.S. government space might make upward of $150,000. Your compensation will depend upon your location, the company, whether you hold a security clearance, or other factors.

What other careers can you pursue?

A career as a SOC Analyst builds a solid foundational skill set and can help you prepare for many cybersecurity careers. Some examples include:

- Cybersecurity engineer
- Incident responder
- Malware reverse engineer
- GRC analyst
- Security architect
- Application security engineer
- Security operations manager

Common interview questions for a SOC Analyst career

In the following sections, you will learn about common interview questions, including general knowledge, attack types, and tools, that you might experience in interviews for a SOC Analyst position. The questions are listed with answers.

 A key item to note is that you want to keep your answers as short as possible during the interview, and then just ask the interviewer whether they need you to expand upon the subject.

Remember that *clear and concise makes the interview nice.*

General SOC knowledge questions

In this section, you will see some general SOC knowledge questions that might be asked in a SOC Analyst interview.

What is the CIA triad?

The CIA triad can be defined as follows:

*   **Confidentiality** is just making sure that only the right people, systems, or applications can access data. Think of confidentiality as locking your data in a safe and only giving access to people you trust.
*   **Integrity** is making sure that data has not been altered.
*   **Availability** is making sure the right users can access the right information when they need to. In some industries, such as critical infrastructure, availability comes before confidentiality and integrity. An example of this would be a nuclear power plant that needs to remain available.

Let's explore an example of the CIA triad being used for a hospital.

Confidentiality would ensure that only authorized users, systems, or applications could access sensitive patient data. This could be accomplished by implementing **multi-factor authentication (MFA)** and **role-based access control (RBAC)**.

Integrity would ensure that a patient's data is unaltered. This could be accomplished by using digital signatures on documentation and by keeping audit logs of any changes to a patient's medical record.

Availability would ensure that a healthcare professional could access the patient's chart during treatment. This could be accomplished by having data backups and building cyber resiliency in the hospital's infrastructure to minimize downtime.

What is information security and how is it achieved?

Information security just means protecting the **confidentiality**, **integrity**, and **availability** (**CIA**) of information.

Information security is achieved through risk management, where you identify valuable information, any assets related to that information, vulnerabilities, threats to the CIA of the information, and the impact on the information and the organization if an incident occurs.

Explain risk, vulnerability, and threat.

A **vulnerability** is a weakness in a system, application, or network that could be exploited by a threat actor. Outdated software on your computer is an example of a vulnerability.

A **threat** is anything that can potentially exploit the vulnerability and cause damage. Using the outdated software example, a threat could be a malicious hacker trying to exploit the vulnerability to steal patient data. Threats can be intentional, like a malicious hacker, or they can be unintentional, like a user accidentally deleting important data.

A **risk** is the likelihood that a threat will exploit a vulnerability and the impact of that event. Using the outdated software example again, the risk would depend upon the type of data being stored on the system, the other security controls put in place to protect the data, and the impact of the data being stolen, leaked, corrupted, or deleted.

What is the difference between asymmetric and symmetric encryption, and which one is better?

Symmetric encryption uses the same key to encrypt and decrypt. Asymmetric encryption uses different keys to encrypt and decrypt.

Both have benefits and drawbacks. Symmetric encryption is normally faster than asymmetric, but the key needs to be transferred over an unencrypted channel. Asymmetric is slower and can be computationally more expensive. It's often good to use a hybrid approach, combining symmetric and asymmetric encryption because it offers secure key exchange and faster encryption for the actual data being transmitted.

What is an IPS and how does it differ from an IDS?

An **intrusion detection system** (**IDS**) detects an intrusion and then will just alert the administrator for them to take further action.

An **intrusion protection system** (**IPS**) will detect the intrusion and then take action to prevent it.

What is a security misconfiguration?

A security misconfiguration is a vulnerability that arises due to incorrect settings, or a lack of proper security protocols being implemented within a system, application, or network.

Some examples of misconfigurations include:

- Keeping default settings, like the default username and password on a router
- Insecure cloud storage settings
- Unpatched software

What is a firewall?

A firewall is like a gate guard. Based on a set of predefined rules, it allows traffic through or not, like a gate guard, allowing you to go through the gate and visit Oprah.

In modern networks, firewalls are still used, but there is really no *perimeter* anymore due to things such as **bring your own device** (**BYOD**).

The world has recently been hit by an attack (e.g., SolarWinds). What would you do to protect your organization as a security professional?

If you have some experience, you can answer this using that experience as a specific example. If this is your first cyber role, then here is an example of how you could answer this question, focusing on the **Observe, Orient, Decide, Act** (**OODA**) loop as part of IR. It's also a good idea to review the IR steps listed in **NIST SP 800-61**.

 To read more about NIST SP 800-61, please check out `https://csrc.nist.gov/pubs/sp/800/61/r2/final`.

OODA loop IR example

- **Observe**

  Upon notification of the SolarWinds attack, the incident responder would analyze data from various security tools, including SIEM and **extended endpoint detection and response (XDR)**, to identify suspicious activity on our network and hosts.

  We'd also analyze the details of the SolarWinds attack being reported in the media, including targeted vulnerabilities, attack vectors, and **indicators of compromise (IOCs)**. This would include collaborating with industry peers and aggregating and analyzing data from threat intelligence feeds.

- **Orient**

  Based on our analysis of the data, we will determine the potential impact on our organization.

- **Decide**

  Analysis of the data we have that shows an incident would trigger the containment strategy of our IR plan. This would help us isolate any affected systems and prevent the attack from spreading further. On the affected systems, we would attempt to gather forensic data, including IOCs. This helps us gather data that can be used to help protect against future attacks. We would then focus on eradicating the attack. This could include rebuilding or reprovisioning the entire system, remediating malware on endpoints that we can remove, and patching vulnerabilities that were exploited.

- **Act**

  Our actions would include implementing our incident response plan, communicating with stakeholders about the incident, and conducting a post-mortem analysis of the incident to identify areas for improvement.

What is the difference between a security policy and a security procedure?

Security policies define the overall security goals and expectations for the organization. They provide clear directions on how employees and other stakeholders should handle sensitive information and use technology resources securely. An **Acceptable Use Policy (AUP)** is an example, and it might state that employees are prohibited from using personal email accounts for work purposes or storing sensitive data on unauthorized cloud storage platforms.

Cybersecurity procedures are detailed, step-by-step instructions that outline how to implement security policies. They provide a practical guide for employees on how to perform specific security-related tasks. A password management procedure might detail specific requirements for password complexity, rotation frequency, and proper storage practices. It would explain how employees should create strong passwords, change them regularly, and avoid writing them down in easily accessible locations.

What is non-repudiation (as it applies to IT security)?

Non-repudiation basically means that neither the sender nor receiver of the information can deny that they processed the information. The sender or receiver could be human-to-human communication, human-to-machine, or machine-to-machine.

What is the relationship between information security and data availability?

Information security entails protecting data and ensuring that only authorized entities can access the data. Data availability just means that the authorized entities can access the data when they need to.

Can you give me an example of a security vulnerability?

For this question, you could focus on items listed in the OWASP Top 10. For example, injection vulnerabilities can occur when an application fails to properly sanitize user input before processing it. An example of an injection vulnerability is a website login form that allows malicious code to be injected, retrieving usernames and passwords from the database.

 To read more about the OWASP Top 10, please check out `https://owasp.org/www-project-top-ten/`.

What is a security control?

Security controls are safeguards, parameters, and countermeasures used to protect data, services, and business operations.

What is information security governance?

Information security governance (**ISG**) is the framework that defines how an organization manages its information security risks. It establishes the processes, structures, and oversight needed to ensure that information security strategies are aligned with business objectives and applicable regulations and laws.

What is risk appetite?

Risk appetite refers to the level of risk an organization is willing to accept in pursuit of its business objectives. It's a calculated decision based on factors like the value of information assets, industry regulations and compliance requirements, the cost of security controls, and the organization's security culture.

What are the risks of open-source software and proprietary software to supply chain security?

- **Open-source software**: Open-source software means that source code is publicly available and anyone can view it. An advantage of this is that a large community can check the code for vulnerabilities and correct issues. However, there are some disadvantages of using open-source software, including the source code being made public, which means a threat actor can identify new vulnerabilities; the speed of applying updates for vulnerabilities can vary based on the open-source project managers, which can leave organizations exposed; and open-source projects often have dependencies from other open-source projects, which means vulnerabilities in those dependencies introduce additional risk.

- **Proprietary software**: Source code is not visible for proprietary software, and vendors will patch vulnerabilities that are discovered in the code. Think of how Microsoft offers update patches to fix issues in the Windows operating system. Proprietary software does introduce some risks, including reliance on vendors to provide patches, dependencies on third-party libraries, and, potentially, an increase in the amount of time before a vulnerability is discovered because a limited number of people analyze the source code.

What is the chain of custody?

The chain of custody is essentially the paper trail showing who has handled evidence from the time it was collected until the time it is presented in a court of law.

What is the Cyber Kill Chain®?

The Cyber Kill Chain® is a framework developed by Lockheed Martin to describe the stages of a cyberattack. It provides a structured approach to understanding and defending against cyber threats by breaking down the attack process into distinct phases.

Here are the seven stages of the Cyber Kill Chain®:

- **Reconnaissance** – in this stage, the attacker gathers information about the target to identify vulnerabilities and plan the attack. This can involve scanning networks, social engineering, and researching public information.

- **Weaponization** – in this stage, the attacker creates a malicious payload, such as malware or an exploit, to deliver to the target. This often involves combining the payload with a delivery mechanism, like an email attachment or a compromised website.

- **Delivery** – in this stage, the attacker transmits the malicious payload to the target using methods such as phishing emails, drive-by downloads, or USB drives.

- **Exploitation** – in this stage, the malicious payload exploits a vulnerability in the target system, allowing the attacker to gain initial access.

- **Installation** – in this stage, the attacker installs malware on the compromised system to maintain access and establish a foothold in the organization.

- **Command and control (C2)** – in this stage, the attacker establishes a command-and-control channel to communicate with the compromised system and issue commands. This could include downloading malware from a remote server.

- **Actions on objectives** – in this stage, the attacker performs their intended actions, such as data exfiltration, system sabotage, or further propagation of the attack.

What is a honeypot?

Honeypots are decoy systems or networks designed to lure attackers and detect malicious activities. They appear as legitimate targets but are isolated and monitored environments. Honeypots are used to gather intelligence on the **tactics**, **techniques**, and **procedures** (TTPs) of threat actors.

What information security challenges are faced in a cloud computing environment?

There are many challenges. A few you should answer with are IAM, security misconfigurations, visibility into your cloud infrastructure and assets, and insider threats.

How many bits do you need for an IPv4 subnet mask?

Subnet masks are 32-bit for IPv4.

What are the layers of the OSI model?

I've listed the layers in the list that follows, but you will also want to understand how data flows through these layers and what the term **encapsulation** means. I've listed encapsulation as the next question, but typically, OSI and then encapsulation will be asked about concurrently in a real job interview:

- **Layer 1** – the **physical layer**, which is where raw bitstream is transferred over a physical medium (that is, fiber optic cable, copper cables, and electromagnetic waves).

- **Layer 2** – the **data link layer**, which controls the transfer of data between nodes on the same LAN segment and contains the sub-layers of **media access control** (**MAC**) and **logical link control** (**LLC**). This layer is where you see the MAC address (for example, `ff:ff:ff:ff:ff:ff`), and the information at this layer is labeled as `frames`.

- **Layer 3** – the **network layer**, which decides what path the data will take. This layer transports and routes the packets across network boundaries. Information at this layer is labeled as a packet, and this is where IP routing lives. An example of an **Internet Protocol version 4** (**IPv4**) address at this layer would be `192.168.0.55`, and an example of an **Internet Protocol version 6** (**IPv6**) address at this layer would be `2001:0DB6:AC10:FE01:0000:0000:0000:0000` or, written in the shorter version, `2001:0DB6:AC10:FE01::::`. Some of the protocols at this layer are **Address Resolution Protocol** (**ARP**), **Reverse Address Resolution Protocol** (**RARP**), **Domain Name System** (**DNS**), **Internet Control Message Protocol** (**ICMP**), and **Dynamic Host Configuration Protocol** (**DHCP**).

- **Layer 4** – the **transport layer**, which transmits data using protocols such as **Transmission Control Protocol** (**TCP**) and **User Datagram Protocol** (**UDP**). The transport layer is responsible for segmenting data from applications into a manageable size, and the information is labeled as a segment at this layer. The UDP protocol is faster than TCP, but it just sends the data and doesn't care whether it was received at the other end. With TCP, a three-way handshake is established, which allows the sender to know whether the data was received by the intended recipient.

- **Layer 5** – the **session layer**, which maintains connections and controls the ports and sessions. The session layer handles the creation, use, and breakdown of a session. It also handles token management for the session.

- **Layer 6** – the **presentation layer**, where data is presented in a usable format and encrypted. This layer preserves the syntax of the data being transmitted and handles compression and decompression.

- **Layer 7** – the **application layer**, where interaction with applications occurs. Some examples of the protocols at layer 7 are **HyperText Transfer Protocol** (**HTTP**), **Secure Shell** (**SSH**), **File Transport Protocol** (**FTP**), and **Simple Mail Transfer Protocol** (**SMTP**).

What is encapsulation?

Encapsulation and encryption are sometimes confused because both involve modifying data for network communication. Encryption is used to protect the confidentiality of the data itself, while encapsulation ensures proper routing and delivery of data packets.

Encapsulation is the process of adding header information to a piece of data as it travels through different layers of a network model (like the OSI or TCP/IP model). The header information acts like a label, specifying the origin, destination, and type of data being sent. The process of encapsulation ensures that data reaches its intended recipient and allows different network layers to handle the data appropriately.

What are the three ways to authenticate a person?

Three main ways to authenticate a person in cybersecurity are something you know (e.g., password), something you have (e.g., 2FA), and something you are (e.g., biometric).

MFA for logging into your bank's app is an example of using these three factors. Your bank requires you to log in with a username and password (something you know), then you receive a one-time code (something you have) to enter, and then the banking app requires you to also scan your face (something you are) before you can log in to your account.

What are false positives and false negatives in firewall detection?

A false negative occurs when a firewall fails to detect and block malicious traffic, allowing a security threat to pass through undetected. A primary risk is that an undetected threat could lead to a cyberattack.

A false positive occurs when a firewall incorrectly identifies legitimate traffic as malicious and blocks it. A risk with false positives is lost productivity due to workflow interruptions from the legitimate traffic being blocked.

What is the three-way handshake in networking?

The three-way handshake is a fundamental process used in the **Transmission Control Protocol (TCP)** to establish a reliable connection between a client and a server over a network. It ensures that both parties are synchronized and ready to communicate.

The three parts of the handshake are:

- Synchronize (SYN)
- Synchronize-Acknowledge (SYN-ACK)
- Acknowledge (ACK)

Let's explore the process further:

- **SYN**: The client system begins the process by sending a TCP segment with the SYN flag set. This segment includes an **initial sequence number** (**ISN**), which is a random value chosen by the client to start the sequence of bytes it will send. The SYN segment indicates a request to establish a connection and synchronize the sequence numbers between the client and server.

- **SYN-ACK**: Upon receiving the SYN segment, the server acknowledges receipt by sending back a TCP segment with both the SYN and ACK (SYN-ACK) flags set. The server's segment includes its own ISN and acknowledges the client's ISN by setting the acknowledgment number to the client's ISN plus one. This segment acknowledges the client's request and provides the server's own ISN, indicating its readiness to establish a connection.

- **ACK**: The client responds to the server's SYN-ACK segment by sending an ACK segment. This segment acknowledges the server's ISN by setting the acknowledgment number to the server's ISN plus one. This final segment confirms that the client has received the server's SYN-ACK segment, and the connection is now established.

What are some of the responsibilities of level 1 and 2 SOC analysts?

This question helps the interviewer understand how much you know about the role and its common responsibilities.

Some responsibilities of a level 1 (tier 1) SOC Analyst include monitoring for malicious and anomalous behavior in network and system traffic through tools, such as SIEMs and IDSs, using ticketing systems, and escalating suspicious activity found to level 2 analysts for review.

Level 2 (tier 2) SOC analysts perform triaging of alerts using playbooks. Level 2 analysts may also tune the collection tools to help reduce false positives and use the MITRE ATT&CK framework (https://attack.mitre.org/) to identify security gaps in an organization's defensive posture. At this level, you will also remove malware from end-user systems and write **YARA** (**Yet Another Recursive Acronym**) rules to detect and stop future attacks.

What are the steps to building a SOC?

This is normally a question asked to a senior level 2 or 3 SOC Analyst. A goal here is to see how you would use your knowledge and experience to architect a SOC from the ground up.

Remember that organizations might choose to build an in-house **Security Operations Center** (**SOC**) or outsource to a **managed security service provider** (**MSSP**). Cost, access to expertise, and scalability are some of the reasons an organization might choose to use an MSSP.

The steps to building a SOC include the following:

1. *Develop your SOC strategy*: The key to developing your strategy is to understand the current state of your organization and perform the following:

    -   Assess your existing capabilities.
    -   Delay non-core functions until your core functions are sufficiently mature.
    -   Identify and define business objectives from stakeholders.

2. *Design your SOC solution*:

    -   Choose a few business-critical use cases (for example, a phishing attack).
    -   Define your initial solution based on these use cases.
    -   Consider that your solution must be able to meet the future needs of the organization.

    Remember, a narrow scope will help reduce the time to initial implementation, which will help you achieve results faster.

3. *Create processes, procedures, and training*:

    -   Identify and analyze threats to determine the nature and extent of risk to the organization.
    -   Implement countermeasures to mitigate threat actors and the associated risk.

4. *Prepare your environment before deploying the SOC*:

    -   Ensure SOC staff desktops, laptops, and mobile devices are secured.
    -   Limit remote access for SOC staff (and third parties if applicable).
    -   Require MFA for all accounts.

5. *Implement your solution and leverage technology where applicable*:

    -   Deploy your log management infrastructure.
    -   Onboard your minimum collection of critical data sources.
    -   Deploy your security analytics capabilities.
    -   Deploy your **Security Orchestration, Automation, and Response (SOAR)** solution.
    -   Begin deploying use cases to focus on end-to-end threat detection and response.
    -   Incorporate threat intelligence feeds.
    -   Employ detection engineering.
    -   Incorporate automation.

6. *Implement and test your use cases:*

- Test your use cases.

- Analyze the security and reliability of your security solution.

7. *Maintain and improve your SOC:*

- Tune to improve detection accuracy.

- Add other systems as inputs or outputs.

- Review the SOC, SOC roles, and staff counts.

What is data protection in transit versus data protection at rest?

Data protection in both scenarios involves encrypting the data. As the name implies, data protection in transit just means you are protecting the data from end to end while it's being transmitted. Data at rest just means the data is protected while it is stored.

Data at rest can be protected with **full disk encryption** (**FDE**), which includes BitLocker on Windows and FileVault on macOS. **Transparent data encryption** (**TDE**) can be used to protect data in SQL Server and Oracle databases, and file-level encryption, like **Encrypting File System** (**EFS**) in Windows, can be used to encrypt individual files and folders. Cloud providers, like Amazon's AWS, offer cloud-storage-level encryption, like Amazon S3 server-side encryption.

Data in transit can be protected with **transport layer security** (**TLS**), **Secure Shell** (**SSH**), **Internet Protocol Security** (**IPsec**), **Pretty Good Privacy** (**PGP**), and **Virtual Private Networks** (**VPNs**).

Is it an issue to give all users administrator-level access?

Yes, this is an issue, and you will want to implement the principle of least privilege as part of IAM.

How can you tell whether a remote server is running Windows Internet Information Services (IIS) or Apache?

You can run a simple scan with a tool such as **Nmap** to see what it is running and the version. You could also do banner grabbing.

How often should you perform patch management?

A risk-based approach is important for patch management. Assessing the risk associated with each patch, and the actual risks to an organization if the patch is not applied, allows the organization to prioritize patches based on the potential impact on critical systems and applications.

Regular patching should typically be performed on a consistent schedule, such as monthly or quarterly, depending on an organization's policies and the criticality of the systems involved. Patching should also be done to ensure compliance with applicable laws and regulations. Organizations might choose to implement additional security controls instead of patching. This is commonly seen in critical infrastructure facilities, like a water treatment facility, where the system cannot be taken offline because it impacts human life. In this case, the organization would implement additional layers of security controls to reduce the risk associated with not patching the system.

What is Docker?

Docker uses OS-level virtualization and delivers infrastructure as code through containers. What does this mean? It means you can run a virtualized infrastructure at low or no cost on just about any computer you have. What does this mean for a company? It usually means significant infrastructure savings.

What is the difference between Transmission Control Protocol (TCP) and User Datagram Protocol (UDP)?

TCP is a connection-oriented protocol, which means it establishes a connection between the sender and receiver before data can be sent. This process involves the three-way handshake (SYN, SYN-ACK, and ACK) discussed earlier to ensure both parties are ready to communicate.

UDP is a connectionless protocol, meaning it does not establish a connection before data transmission. Data is sent without prior agreement between the sender and receiver. UDP does not guarantee delivery of the data, but it is faster than UDP. UDP is used to stream videos and online gaming, where a few dropped packets will not impact the overall user experience.

What is a playbook/runbook in SOC?

A playbook, also known as a **standard operating procedure** (**SOP**), consists of a set of guidelines to handle security incidents and alerts in the SOC. For example, if credentials were compromised, the playbook would help the level 1 SOC Analyst know what actions they should take.

What is DNS?

DNS is basically the phone book (I might be giving my age away with this example) of the internet. As an example, let's say that you type google.com in your browser and the domain name (google. com) is then translated to an IP address (192.168.0.1 for this example) for Google's servers, so you can see the information on their website. This eliminates the need for you to memorize every server IP address of Google.

There are four DNS servers involved in your request to access Google's web page:

- The **DNS recursor** receives queries from clients and then makes any additional requests to satisfy the client's DNS query. This is similar to you requesting a book from the library, the librarian looking up the shelf that the book is on, and then handing the book to you.

- The **root nameserver** is the first step in translating human-readable information (that is, google.com) into an IP address. Using our library analogy, this is like the index card that tells you that the book is in the non-fiction section. Using our google.com example, this nameserver would tell you that the web page you want is in the *Google* section of the library.

- The **top-level domain** (TLD) nameserver is the next step and hosts the last portion of the hostname (.com in our google.com example). In our library example, this would be the librarian telling you that the book is on shelf 12 of the *Google* section.

- The **authoritative** nameserver can be thought of as the master index card that tells you specifically where the book is in the library.

  One thing to keep in mind is that DNS uses multiple servers and not a single server.

You receive an email from your bank stating that there is a problem with your account. The email states you need to log in to your account to verify your identity and even provides a link to your bank. If you don't verify your identity, the email states that your account will be frozen. Tomorrow is payday and you need to pay your rent, which is past due, via a wire transfer in the morning. What should you do?

This is a simple phishing attack, and the question is designed to test your general knowledge of phishing, since you might have end users calling the SOC with this issue. In this example, you should not click any links or documents in that email. Instead, visit the bank's website URL directly or call your bank. You should also change the password for your bank account. Most banks don't typically contact you via email if there is an issue with your account.

After answering this question, you can ask the interviewer whether they need you to explain what phishing is.

You are a new level 1 SOC Analyst and receive a call from the IT helpdesk to ensure you can access all systems. The IT helpdesk person is friendly to you and asks you to confirm your password so that they can verify you meet the minimum complexity requirements. What do you do?

This is a **vishing (phishing via phone)** attack. One part of this answer is to hang up the phone, but before you do, I would try to get as much information as possible from the individual. I've had this happen before and I was able to get a postal mailing address (I told them I wanted to personally mail them a thank you card for being so helpful) out of the individual.

The mailing address turned out to be a mailbox at a UPS store, but it was an additional clue for law enforcement to hopefully catch the criminal calling me.

What is cognitive cybersecurity?

Cognitive cybersecurity is the application of **artificial intelligence (AI)**, patterned on human thought processes, to detect threats and protect physical and digital systems. It uses data mining, pattern recognition, and natural language processing to simulate the human brain.

What is the difference between SIEM and IDS systems?

SIEM and IDS systems collect log data.

SIEM tools facilitate event correlation to identify patterns that might indicate an attack has occurred and centralize log data.

IDS tools also capture log data but do not facilitate event correlation. An IDS detects an intrusion and alerts it.

What is port blocking?

The answer is simply blocking ports. It's helpful to block unnecessary ports so that you can reduce the attack surface. One thing to keep in mind though is that many threat actors just use ports they know will always be open (HTTPS on port 443 as an example).

What is ARP and how does it work?

Address Resolution Protocol (ARP) is a protocol for mapping an **Internet Protocol address (IP address)** to a physical machine address that is recognized in the local network.

How does it work?

1. When an incoming packet is destined for a host machine on the LAN at the gateway, the gateway asks the ARP program to find the physical host or MAC address that matches the IP address.

2. The ARP program looks in the ARP cache and, if it finds the address, provides it so that the packet can be converted to the right packet length and format and sent to the host machine.

3. If no entry is found for the IP address, ARP broadcasts a request packet in a special format to all the machines on the LAN to see whether one machine knows that it has that IP address associated with it.

What is Address Resolution Protocol (ARP) poisoning?

ARP poisoning, also known as ARP spoofing, is a type of cyberattack where an attacker sends falsified ARP messages over a local network. The aim is to associate the attacker's **Media Access Control (MAC)** address with the IP address of a legitimate device on the network, such as a gateway or a server. This allows the attacker to intercept, modify, or disrupt traffic between devices on the network.

What is port scanning?

Port scanning is a technique used to identify open ports and services available on a host. A threat actor can use port scanning to identify services running and identify vulnerabilities that can be exploited. A network administrator might use port scanning to verify the security policies set on the network. **Nmap** is a popular tool that can be used for port scanning.

A senior executive approaches you and demands that you break security policy to let them access a social media website. What do you do?

In this situation, I would ask why they need access and then explain that it is against the security policy. If the executive persists, I suggest getting your leadership team involved. Many companies have a formal process for one-off requests like this to be reviewed and approved.

What is an insider threat?

Insider threats in cybersecurity refer to risks posed by individuals within an organization who have access to its systems and data. These threats can be categorized into intended (malicious) and unintended (accidental) threats.

- **Intended (malicious) insider threats**

  Intended insider threats involve deliberate actions by individuals with authorized access who aim to harm an organization for personal gain, revenge, or other malicious reasons. These threats are often motivated by financial incentives, revenge, ideological beliefs, or coercion.

Examples include:

- Data theft
- Sabotage
- Espionage
- Fraud

- **Unintended (accidental) insider threats**

  Unintended insider threats occur when employees or other authorized individuals inadvertently cause harm to an organization due to negligence, lack of awareness, or simple mistakes. A key point is that the actions are not deliberate and usually result from human error, ignorance, or lack of training.

  Examples include:

  - Phishing attacks
  - Misconfiguration
  - Data leaks
  - Weak password practices
  - Lost devices

What is a residual risk?

Residual risk in cybersecurity refers to the amount of risk that remains after all mitigation efforts have been applied. It's important to understand an organization's risk appetite, which is the level of risk the organization is willing to accept. Once the residual risk is identified, organizations can choose to accept the remaining risk, implement additional security controls (e.g., encryption), or transfer the risk (e.g., cyber insurance). The organization also needs to continuously monitor security controls to ensure they are still reducing risk as expected and as the organization's risk appetite changes.

What is Data Loss Prevention (DLP)?

DLP tools are used to ensure that users do not send sensitive data outside an internal network.

Best practices for DLP include identifying data, classifying it, prioritizing it, understanding the risks to the data, monitoring data in transit, and creating controls to protect the data. You will also want to train your employees because many will not understand how their actions can result in data loss.

What is an Incident Response Plan (IRP)?

IR plans ensure that the right people and procedures are in place to deal with threats. This allows your IR team to perform a structured investigation into events to determine the **indicator of compromise (IOC)** and the **tactics, techniques, and procedures (TTPs)** of the threat actor(s). An IR plan is like a step-by-step guide to follow if an incident occurs; however, you might jump around through different phases of the Kill Chain depending on the incident.

NIST 800-61 is a good resource for you to learn about the different phases of incident handling, and you will likely be asked some questions relating to 800-61 in job interviews.

I've listed the phases as follows:

1. Preparation
2. Detection and analysis
3. Containment, eradication, and recovery
4. Post-incident activity

General attack knowledge questions

In this section, you will see some of the attack knowledge questions that might be asked in a SOC Analyst interview.

What is a botnet?

A botnet is composed of hijacked computers that are used to perform several tasks, including attacks such as DDoS. Some notable botnet infrastructures are Mirai, which hijacked IoT devices, and Emotet.

What are the most common types of attacks that threaten enterprise data security?

The answer to this will change as time progresses and new threats emerge, but in general, it includes things such as malware/ransomware, DDoS/DoS attacks, phishing/**business email compromise (BEC)**, credential stuffing, and web application attacks. Threat actors also use generative AI to build more sophisticated phishing attacks.

The Verizon **Data Breach Investigations Report (DBIR)** is a good source of information for the most prevalent attacks.

To read more about DBIR, please check out `https://www.verizon.com/business/` `resources/reports/dbir/`.

What is XSS and how can you mitigate it?

Cross-site scripting (XSS) is a JavaScript vulnerability in different web applications. There are different types of XSS, including reflected and stored XSS. For reflected XSS, a user enters a script on the client side, and this input gets processed without being validated. This means the untrusted input is executed on the client side, typically through the browser.

For stored XSS, a malicious script is injected directly into a vulnerable web application and executed. This means any user visiting the web app server will be infected, even if they clear their browser cache.

SOC tool questions

In this section, you will see questions on common SOC tools. Please note that this list will not contain every tool and that it's more important for you to understand what the different types of tools used in a SOC are, versus knowing how to use every vendor tool. It's also important to note that the answers to these questions may change as tools are updated with new features.

What is a SIEM?

Security Information and Event Management (SIEM) collects data from various network and host tools and analyzes the data for suspicious activity. Some SIEM tools leverage AI for predictive analysis and response, which helps reduce alert fatigue for analysts.

Why is Splunk used for analyzing data?

It offers business insights, which means it understands patterns hidden within data and turns them into real-time business insights that can be used to make informed business decisions. This is key because there is so much data to sift through in a typical enterprise, and it's important to gain actionable insights into the data. It also provides visibility into your operations and proactive monitoring.

Your answer here should be clear and concise on a few of the value props of Splunk or another SIEM tool. You can always ask the interviewer whether they would like you to provide more context or information.

What do Security Orchestration, Automation, and Response (SOAR) solutions provide that SIEM tools usually don't?

SOAR solutions provide several capabilities that typically extend beyond what SIEM tools offer.

This includes IR playbook automation, orchestration of multiple security tools, incident handling capabilities, advanced analytics with **machine learning** (**ML**) algorithms that can help identify patterns and predict threats, and integration of threat intelligence to provide more contextual information on threats.

Which of the following uses a user's behavior as part of their process to determine anomalous behavior on a network?

- EDR tools
- SIEM tools
- SOAR tools
- UEBA tools

The answer is UEBA tools. **UEBA** stands for **user and entity behavior analytics**. These tools are used to detect attacks faster by aggregating data from on-premises, the cloud, and multiple devices to detect anomalous behavior on a network that might be seen when an attacker moves into lateral movement.

Which of the components listed are seen in many next-gen SIEM solutions but are not traditional SIEMs?

- **Threat intelligence feed**
- **EDR**
- **SOAR**
- **UEBA**

UEBA and SOAR are often seen in next-generation SIEM solutions.

Select all of the SIEM tools from the following:

- Splunk
- QRadar
- Cisco ASA
- Microsoft Sentinel

The only one listed that is not a SIEM tool is Cisco ASA, which is a firewall.

As you can see, many interview questions as a SOC Analyst are around attack types and fundamental knowledge of SIEM tools. The good news is that many companies are only looking for you to have knowledge of attacks and how threat actors might attack the organization as a tier 1 SOC Analyst (entry-level). I would also suggest you explore the MITRE ATT&CK framework to think through how organizations can use it operationally to build detection logic in their SIEM tool, and how it can be used strategically by organizations to identify gaps in their security postures.

Summary

In this chapter, you learned about the SOC Analyst career and the average salary range in the United States. You also learned how this can be a stepping stone into other cybersecurity careers, and you learned about common interview questions asked for SOC Analyst roles.

In the next chapter, you will learn about a career as a penetration tester, including common knowledge-based interview questions you might be asked.

Join us on Discord!

Read this book alongside other users. Ask questions, provide solutions to other readers, and much more.

Scan the QR code or visit the link to join the community.

`https://packt.link/SecNet`

4

Penetration Tester

In this chapter, you will learn what a **Penetration Tester** (**pentester**) is and the average salary range for this career in the US. You will also learn about career progression options and learn common interview questions for the role.

The following topics will be covered in this chapter:

- What is a pentester?
- How much can you make in this role?
- What other careers can you pursue?
- Common interview questions for a pentester career

What is a Pentester?

Penetration testing (**pentesting**), or ethical hacking, is where you assess the security of networks, websites, endpoints, mobile devices, wireless devices, **operational technology/industrial control system** (**OT/ICS**) infrastructure, and the security of physical facilities. This assessment might include performing vulnerability scanning and analysis, reviewing source code, performing **open source intelligence** (**OSINT**), gaining access to a target by exploiting vulnerabilities, escalating privileges, maintaining persistence, and more.

A key thing here is that you have permission as a pentester to attack the target, as defined in the **statement of work (SOW)** of the **penetration test (pentest)**. If you don't have permission, then it's illegal. Before starting any pentest, you need to review the **rules of engagement (ROEs)**, determine the scope of the pentest, and verify that the client owns everything that's listed. I've reviewed SOWs before where the client mistyped an **Internet Protocol (IP)** address, and we could have been in legal trouble for performing the pentest if we had not corrected the documentation. The right documentation is like a get-out-of-jail-free card during a pentest.

The goal of pentests is to simulate which vulnerabilities are exploitable by an adversary, and this is where vulnerability assessments and pentests differ. A vulnerability assessment just identifies that there might be something an adversary can exploit and recommends mitigation, and a pentest shows that it can be exploited and provides ways to mitigate the impact.

A good way to think of the difference between vulnerability assessments and pentesting is a car. Your mechanic runs a diagnostic scan (this is the vulnerability assessment) on your car and identifies some error codes that tell the mechanic five problems that might be the cause. The mechanic then tinkers under the hood to manually assess those potential problems (this is the pentest), and ultimately determines the root cause. For example, you turn on the ignition on your car, but it just will not catch and start the car. This could be caused by the battery, the ignition switch, spark plugs, or other parts. Your mechanic runs a diagnostic scan that will indicate all of these as potential issues, checks each one to identify what the real problems are, and provides recommendations to fix them.

There are many different areas of pentesting that you can specialize in, including applications (such as web apps, cloud, thick clients, and mobile apps), infrastructure/networking, ICS, physical, red team, hardware, **Internet of Things (IoT)**, and social engineering. Many pentesters specialize in one or two of these areas and also have knowledge and skills in other areas. No one is an expert in every area of pentesting, contrary to what you might see in the movies. Speaking of movies, real pentesting has nothing to do with wearing a hoodie in your mom's basement as binary code scrolls across the computer screen. Real-life pentesting takes careful planning and doesn't always involve you being an expert in computer programming, but it can be challenging and rewarding.

So, what skills do you need to be a pentester? For soft skills, passion and the ability to communicate the results of your pentest to stakeholders are critical. For technical skills, you need to have a solid foundation in operating systems, networking, and security.

You also need to be able to use common penetration testing tools, which could include **Burp Suite**, **Metasploit**, **Nmap**, **Exploit Pack**, or other tools used by your organization.

The good news is that, as with most cybersecurity careers, you don't need a college degree or certifications to become a pentester.

If you are looking to gain hands-on experience with home labs, you can download Virtual-Box (`https://www.virtualbox.org/`) or VMware Workstation (`https://www.vmware.com/products/workstation-player/workstation-player-evaluation.html`) for free and install Kali Linux (`https://www.kali.org/`) and Metasploitable (`https://sourceforge.net/projects/metasploitable/`) to practice.

You can also find free Microsoft Windows (`https://www.microsoft.com/en-us/evalcenter/`) **International Organization for Standardization** (**ISO**) images here to build Windows **virtual machines** (**VMs**).

Heath Adams, who is a professional pentester, also has free ethical hacking training videos on YouTube. PortSwigger (`https://portswigger.net/web-security`) also has some free training for web application pentesting. If you just do a quick search online for *ethical hacking training* or *penetration testing training*, you should find hundreds of free and low-cost resources to help build your skills.

How much can you make in this role?

The salary range for a pentester in the US depends on a number of factors, such as your location, the size of the company you work for, certifications you hold, experience, education, and your skills. I've seen salaries as low as $67,500 for junior penetation testers and as high as $270,000 for specialized public sector work. For a junior-level pentester, you can usually expect between $67,000 and $100,000, depending on the factors I mentioned before. I do want to mention that there are far more jobs available on the defensive side than on the offensive.

What other careers can you pursue?

A career as a pentester means you have mastered certain technical and soft skills, so it can help prepare you for any new roles in the industry. I've typically seen pentesters move into other types of pentesting (that is, application instead of infrastructure) or move into leadership roles in the C-suite. Your penetration testing skills can also be applied to careers, like threat hunter, red team specialist, security consultant, incident responder, and security architect.

Common interview questions for a pentester career

The questions that follow are primarily knowledge-based questions. During a junior pentester interview, you will likely experience many knowledge-based questions, and possibly some hands-on testing assessments. In senior and principal pentester job interviews, you often receive a hands-on test of your pentesting skills after the initial phone screen from the recruiter or **human resources** (**HR**). You're likely to encounter questions similar to the following.

Where do you go to research the latest vulnerabilities, and why?

Example answer:

Your answer could include following specific security researchers on Twitter, following blogs such as Krebs and Threatpost, podcasts you listen to, and more. There isn't usually a wrong answer here, but the interviewer does want to see how you stay current on recent vulnerabilities and the latest cybersecurity news.

Some blog websites you might want to check out include PortSwigger's Web Security Blog (`https://portswigger.net/blog`), the Hacker One blog (`https://www.hackerone.com/vulnerability-and-security-testing-blog`), and the Rapid7 blog (`https://www.rapid7.com/blog/`).

What are some areas you are planning to improve in?

Example answer:

This question is being asked to see whether you are a continuous learner and to see how you identify areas of self-improvement. Even as a junior pentester, you should expect to be learning something new continuously, and you need to be able to assess your skill set and know the areas you need to improve in. For example, I'm good at social engineering but not so good at programming. As a pentester, I focused less practice on social engineering since that came naturally and focused instead on becoming better at coding so that I could write my own tools.

Walk me through your process for performing an internal penetration test

Example answer:

The high-level process you might use for conducting an internal penetration test is as follows:

1. Planning/scoping
2. Information gathering
3. Analyzing vulnerabilities
4. Exploiting vulnerabilities

5. Post-exploitation (e.g., gathering additional information, escalating privileges, etc.)

6. Reporting on findings, including remediation recommendations, and assisting in reme-
    diation (if applicable)

7. Clean up any artifacts (e.g., shells) from target systems

8. Lesson learned from the test (sometimes called a hotwash or after-action report)

You discover a SQL injection vulnerability in a web application. How would you exploit this vulnerability, and what steps would you take to secure it?

Example answer:

First, I would use a tool like SQL map (`https://sqlmap.org/`) to automate the exploitation of the SQL injection to understand the extent of the vulnerability. Then, I would manually test different payloads to see if I can extract sensitive data. To secure it, I would recommend parameterized queries or prepared statements, proper input validation, and server-side sanitization.

How can you perform XSS if <script> or alert tags are blocked?

Example answer:

If `<script>` tags are blocked, you could use things such as image payloads or video payloads. Instead of using `alert` tags, you could use tags such as `prompt` and `confirm`.

You need to perform a black box penetration test on a client's network. How would you approach this task?

Example answer:

I would start by verifying the ROE and other legal documentation, such as **service-level agreements (SLAs)**, are in place. I would then begin reconnaissance to gather information about the target network to identify vulnerabilities and then try to exploit any found vulnerabilities. I would then report my findings to the client, along with recommendations for mitigation.

During a penetration test, you find an open RDP port on a server that is within the scope of the ROE. What would be your next steps?

Example answer:

I would attempt to connect to the RDP service using a tool like **rdesktop** or **xfreerdp** to see if there are weak or default credentials that I can exploit. I would also check for known vulnerabilities in the RDP service, such as **BlueKeep**, and attempt to exploit them if they are unpatched.

I would then document my findings, along with recommended mitigations, and report this information to the client.

You have been asked to perform a social engineering test. Walk me through how you would conduct this test.

Example answer:

I would start by researching the target organization using OSINT sources, including social media business pages and employee profiles. This data will help me conduct the social engineering attack through social, email, SMS, vishing, or other method of gaining initial access. I would document my findings, along with information on which attacks are the most successful, and report this information to the client along with recommendations for mitigation.

A client asks you to test the security of their mobile application. What are some of the tools and techniques you would use?

Example answer:

After validating that the legal paperwork is in order, I would use tools such as Burp Suite to intercept and analyze the application's traffic, **APKTool** or **JADX** to decompile the app for static analysis, and **MobSF** for automated security assessment. I would test for common vulnerabilities such as insecure data storage, weak encryption, and improper session handling. I would report the findings to the client, along with mitigation strategies.

You discover an insecure direct object reference (IDOR) vulnerability. How would you exploit this vulnerability, and what are some recommended mitigation strategies?

Example answer:

I would start by identifying which parameter is vulnerable to IDOR. For example, this could be found in a URL, a form field, or an API request where a user-supplied value directly references a resource (e.g., a user ID or a file name).

I would then test different values. For example, changing a user value in a URL from 138 to 139 might allow me to see the data for user 139. I can use tools such as **Burp Suite** to automate this testing process. I would then analyze the results and determine which resources are accessible, the type of data that was exposed, and the potential impact of the vulnerabilities.

Mitigation strategies could include object validation, using indirect references, implementing proper access control, and continuous monitoring and auditing to identify these vulnerabilities in the future.

You find a misconfigured S3 bucket with sensitive data exposed during a penetration testing engagement. What are some of the actions you would take and why?

Example answer:

- Notify the client immediately about the misconfiguration and the potential exposure of sensitive data. This ensures the client is aware of the risk and can act to mitigate it.

- Advise the client to immediately restrict access to the S3 bucket using appropriate **identity and access management (IAM)** policies. Ensure that the bucket is not publicly accessible, and that access is limited to authorized users only.

- Make sure that the S3 bucket's settings are configured to disable public access. This can protect against unauthorized access to the data.

- Suggest encrypting the data stored in the S3 bucket to protect it from unauthorized access, even if the bucket is misconfigured in the future. AWS provides server-side encryption options that can be easily enabled.

- Enable logging and monitoring on the S3 bucket to keep track of access attempts and actions performed on the data. This can help in identifying and responding to suspicious activities.

- Recommend conducting regular security audits and reviews of the S3 bucket configurations and access policies to ensure ongoing compliance with best practices.

- Encourage the client to educate and train their staff on cloud security best practices, focusing on the importance of proper configuration and access control.

A client wants you to perform a physical penetration test. Walk me through your process for this.

Example answer:

I would start by verifying all legal paperwork was in order and verify the scope and ROE of the physical penetration test. I would then gather information on the target (on site if possible), specifically looking for weaknesses I could exploit. For example, if the facility gets a delivery from FedEx every Saturday and there is a different driver each time, this could be my way into the facility.

Other sources of information include OSINT and building schematics. I would then identify my way in and execute my plan to hit my objective. The objective is determined by the ROE and might include accessing a specific part of the facility or even accessing a server room. I would then document and report my findings to the client, along with recommendations for improving security.

You are conducting a penetration test and gain access to a sensitive database. What are your ethical responsibilities and next steps?

Example answer:

My ethical responsibility is to protect the client's data. I would document the access method and any sensitive information accessed, then immediately inform the client. I would also provide recommendations for securing the sensitive data.

During a penetration test, you encounter an unfamiliar service running on a non-standard port. How would you proceed?

Example answer:

- I would use a tool such as **Nmap** to perform a detailed scan of the port to identify the service and version running on it. An example command would look like `nmap -sV -p [port] [target]`.
- I could also use a tool such as **Netcat** to grab the service banner, which can provide clues about the running service and version.
- Once I have more insight into the service running, I can search for known vulnerabilities associated with the identified service using databases such as **CVE**, **Exploit-DB**, and vendor advisories.
- Looking at vendor documentation and technical manuals for the service can also be helpful for understanding its functionality, configuration, and common weaknesses.
- Once I identify potential vulnerabilities and possible exploits, I can automate exploitation with a tool such as **Metasploit**.
- I would then document my findings and report the information to the client, along with recommendations for mitigation.

How do you scope out a penetration testing engagement?

Example answer:

- Define the objectives, including what assets, systems, or data need to be tested and what specific threats or vulnerabilities are of concern.

- Identify the scope of the engagement, including which networks, applications, devices, and systems are in scope and which are out of scope. This includes specifying IP addresses, domains, and physical locations.

- Decide on the type of penetration test (black box, white box, or gray box) based on the information provided by the client and the depth of testing required.

- Identify any constraints or limitations, such as testing windows, restrictions on certain types of tests (e.g., no Denial-of-Service attacks in a cloud environment), and any legal or regulatory considerations.

- Conduct a risk assessment to identify potential risks to the client's operations during the testing process and discuss mitigation strategies to minimize disruption.

- Determine the necessary resources, including tools, personnel, and time required to conduct the penetration test effectively.

- Establish clear communication channels and protocols for reporting findings, coordinating with the client's team, and addressing any issues that arise during the engagement.

- Prepare and sign all necessary documentation, including engagement agreements, **non-disclosure agreements** (**NDAs**), and **scope of work** (**SOW**) documents, to formalize the engagement.

What are some ways you can gather information on a target during a pentest?

Example answer:

Some of the common ways to get information on a target include more passive activities, such as OSINT, and more active techniques, such as running a **Network Mapper** (**Nmap**) scan. Your specific actions will depend on the scope of the pentest. If you get this question in an interview, I suggest asking the interviewer about the scope of the pentest because that will help guide your answer to this question.

What is social engineering?

Example answer:

Social engineering is basically the use of human psychology to influence someone else's behavior.

The components of a successful social engineering attack include an evaluation of the target and their weaknesses, the ability to perform pretexting (where the attacker creates a false scenario to get sensitive data or trick a user into a specific action), the ability to exploit human psychology for the attacker's benefit, the ability to build a perceived relationship with the target, and the ability to get the target to take some sort of desired action.

Here's a simple example of social engineering. You and I are at a coffee shop, and I convince you to buy me a cup of coffee. Perhaps I say that I have left my wallet at home because I'm stressed out that my kid is in hospital, and you feel sorry for me and buy the cup of coffee because you have little kids of your own.

In this example, I'm just getting a cup of coffee, but what if I sent you an email with a malicious GoFundMe link embedded with a keylogger and used the same story about my kid in the hospital? You might click the link to donate, be redirected to the real GoFundMe page, and make a donation to help. Meanwhile, I've dropped malware on your system and now track every keystroke you make as you log in to your bank account to see whether the GoFundMe donation has registered on your account balance.

One thing to keep in mind is that during an interview, you might be asked to conduct a social engineering attack and then continue your (simulated) attack through the organization after gaining initial entry. The next steps after entry can include things such as enumerating user accounts on the system to identify administrator accounts, privilege escalation, network enumeration, deploying ransomware, and enumerating Active Directory with a tool such as Bloodhound (`https://github.com/BloodHoundAD/BloodHound`).

What are some ways to perform physical pentesting?

Example answer:

Before answering this question, it's usually best to start with a short overview of what could happen if physical security were breached. If you breach the physical security of a target, you could steal devices, documents, and data, take photographs or videos of restricted areas or proprietary systems and additional security defenses being used to protect them, and then plant things such as keyloggers (via a **Universal Serial Bus** (**USB**) drop attack) and set up rogue devices on the target's network.

Common physical security controls that are put in place to stop attackers include door locks (physical/electronic), surveillance cameras and security alarms, security guards, perimeter walls and gates, security lights, motion sensors, and human traps that restrict individuals to a specific area.

Physical pentesting can include dumpster diving, lock picking, cloning badges, bypassing motion detectors, jumping fences or walls, bypassing or interrupting the feeds of surveillance, cameras, and **radio-frequency identification** (**RFID**) replay attacks.

What are the types of social engineering?

Example answer:

There are several types of social engineering attacks, including the following:

- **Phishing attacks**: These are typically done via email whereby the attacker is looking to obtain sensitive information or get the recipient to perform a specific action (such as transfering money to a bank account controlled by the attacker). There are several forms of phishing attacks, such as these:

  - **Phishing emails**: These are the most common form of phishing attacks, and you will typically see them done against a broad range of targets—in the case of spam—or more narrowly focused—in the case of **business email compromise (BEC)** attacks. BEC attacks usually involve spear-phishing and whaling. Phishing attacks are the most common entry point of attacks, including ransomware attacks.

  - **Spear-phishing attacks**: These are targeted phishing attacks against a specific person or group. The attacker would need to gain information about the target and craft a message, across any medium, that would entice the victim to take some sort of action.

    An example would be the attacker knowing you love drinking coffee from Starbucks. Through social media posts, the attacker identifies two locations you typically go to and then sends you a coupon link through social media for a free cup of coffee at one of those locations. In one of my training programs, a student was able to get an instructor to click a fake link with a similar type of attack for a free donut. Fortunately for the instructor, this was done in a controlled setting and the link was not really malicious.

    Another example of a spear-phishing attack is the threat actor noticing employees at a company order from the same restaurant at lunch each day and then compromising the restaurant's website with malware so that each employee visiting the website gets their system infected. This is known as a **watering-hole attack**. Watering-hole attacks are often used to target specific industries or specific companies.

  - **Whaling attacks** are another form of a targeted phishing attack. The main difference between whaling attacks and spear-phishing attacks is that a whaling attack focuses on a powerful or wealthy individual, such as the **chief executive officer (CEO)** of a major company. A whaling attack is often harder to pull off successfully, but the financial reward for the attacker could be in the millions.

- **Tailgating** is another social engineering attack where the adversary gains access to a secure area by following an authorized employee inside. In this case, the employee does not know the attacker has followed them in, and this can happen if the employee opens the door wide or if it takes time for the door to close after the authorized employee. This attack is hard to pull off if there are security guards or if the authorized employee is situationally aware.

- **Piggybacking** is an attack whereby the victim is tricked into letting the attacker in. This can happen a lot at large companies, where the attacker mentions they work in a different department and just left their badge at home. Forgetting a badge or other employee ID happens a lot in companies, and many employees would empathize with the attacker and let them inside.

I worked at a healthcare organization where every day, someone would forget their badge to scan in and wait at the door for someone else to let them in. Even back then, I implemented zero trust and would decline to let the person in, even if they worked in my department. My argument was that I didn't know whether HR had fired them last night and they were unauthorized to be in the building. Needless to say, that didn't make me popular with some coworkers, but they did understand my point of view a few months later when a man with a gun was able to gain entry into the building because someone else thought he worked there and had just forgotten his badge.

Some other attacks you might see referenced in certification study material are hoaxes, elicitation, spam, and impersonation. In my experience, these are normally coupled with the previous ones mentioned. For example, a **hoax** is simply where the attacker presents a fictitious situation. An example of this is when you receive a phishing email from your *bank* stating there is an issue with your account, and you need to verify your identity by logging in to your account from a link in the email. If you click the link, you are taken to a fake login page that will capture your username and password.

How can a company protect against social engineering attacks?

Example answer:

Some ways to help protect against social engineering attacks are **two-factor authentication (2FA)**, security awareness training, granular access control, logical controls (such as blocking USB ports on hosts), and proper security policies.

When I did security awareness training for healthcare companies, I would always relate each recommendation to how it impacted the employees' day. For example, I would ask the nursing staff what would happen to their license if they shared their login credentials with me and I went in and altered 90% of their nursing notes on patients.

How would they know which notes I had altered? What would local, state, and federal agencies do to them and their license? How would it impact their patients and the care that they received? When you put training into context for people, they are more likely to follow best practices.

What is the content of a well-written pentest report?

Example answer:

A pentest report is important and should contain the following items:

- A cover page.
- An executive summary should be one page or less and should highlight exciting pieces of the report's findings. Think of this part as marketing, and you need to get the stakeholder to buy what you are selling so that they finish reading the full report.
- A summary of vulnerabilities that you found. A simple pie-chart graphic works well for this if you categorize the vulnerabilities. The vulnerability reporting should include the risk rating (qualitative and quantitative), risk impact and likelihood of the vulnerability being exploited, and severity levels.
- Details of the testing team and tools that were used in the engagement.
- A copy of the original scope of work that was signed as part of the contract. It's helpful to have this in the report as a reference for the client.
- The main body content of the report, which goes into detail in terms of your findings.

How can you identify whether a web application that you came across might be vulnerable to a blind Structured Query Language (SQL) injection attack?

Example answer:

You can use the `sleep` command, and if the web app sleeps for a period of time, it could indicate that it is vulnerable.

What is a MITM attack?

Example answer:

In a **man-in-the-middle** (**MITM**) attack, the attacker acts as a relay between the client and the server. You can use things such as **HyperText Transfer Protocol (HTTP) Strict Transport Security** (**HSTS**) and digital signatures of packets to protect against MITM attacks. Some popular tools for performing MITM attacks are **Wireshark**, **Ettercap**, **Nmap**, **Metasploit**, and **Netcat**.

What is CSRF?

Example answer:

Cross-Site Request Forgery (**CSRF**) attacks take advantage of the trust relationship that is established between the user and a website. The attacker uses stored authentication in browser cookies on the user's side to authenticate to the website. An example is where you have a login to a shopping website and you store the authentication in cookies in your web browser so that each time you visit the shopping website, it authenticates you and takes you into your account. An attacker could craft a **Uniform Resource Locator** (**URL**) with a parameter to increase the number of items added to your shopping cart when you are purchasing an item. You might not notice this and end up purchasing the additional items.

What is an open redirect attack?

Example answer:

In an open redirect attack, the parameter values of the HTTP GET request allow information to be entered that can redirect the user to a different website. The redirect could happen once on the loading of the website page or after the user has taken an action such as logging in to the site.

In this example, the RelayState parameter is not being validated by the website, so an attacker could replace the legitimate website with their malicious code and the user would be redirected to the malicious site.

Correct URL: `https://www.microsoft.com/login.html?RelayState=http%3A%2F%2FMicroso`
`ftGear.com%2Fnext`

Attacker URL: `https://www.microsoft.com/login.html?RelayState=http%3A%2F%2FBadGuy`
`Website.com`

This type of attack is commonly used in phishing emails, where the victim is redirected to a fake login page (for their bank, PayPal, and so on) after clicking a link in the email. After they enter their login credentials, the victim is then redirected to the real website and asked to enter their login credentials again.

What cookie security flags are there?

Example answer:

The HttpOnly flag can be used to block access to the cookie from the client side, which can mitigate XSS attacks.

The `Secure` flag forces cookies to be transported over **HTTP Secure** (**HTTPS**) instead of HTTP.

What is the last pentest tool that you've improved, fixed, and/or contributed to?

Example answer:

This question is aimed at experienced pentesters, and it's designed to help the hiring manager identify how you are giving back to the community.

Can you identify the most common HTTP methods and how they can be used in attacks against web applications?

Example answer:

Common HTTP methods include `GET`, `POST`, `PUT`, `DELETE`, and `TRACE`. `GET` and `POST` are used in attacks by modifying the parameters. An attacker could use `PUT` to upload arbitrary files on the web server. `DELETE` could be used in a **denial-of-service** (**DoS**) attack. `TRACE` could be used to return the entire HTTP request, which would include cookies. An attacker could leverage `TRACE` to perform a **cross-site tracing** (**XST**) attack where the attacker uses XSS to retrieve `HttpOnly` cookies and authorization headers.

What are the differences between attacking a web application and an Application Programming Interface (API)?

Example answer:

Web applications have traditionally involved one request to one server, so you just needed to protect one application. With APIs, you have hundreds of requests to hundreds of microservices, which means you now have to protect hundreds of small applications. The main API security flaws being exploited are around authentication and authorization, and each microservice needs to verify identity and permissions before granting access. A challenge in API security is visibility into your APIs because shadow APIs might exist (those that developers have forgotten about), and if they are public-facing, they can be exploited.

How do you measure the results of a pentest?

Example answer:

It depends on what the organization is looking to measure. Common things to track are the criticality of findings, how many issues that surfaced in the pentest actually get fixed, what types of vulnerabilities and exploits are being discovered, and which new issues have been identified since the last pentest. In addition, client feedback is a good indicator of the success of a penetration test.

How can you leverage threat modeling in a pentest?

Example answer:

Threat modeling helps the pentester identify critical business assets and the impact on the organization if those assets are compromised by an attacker. It also helps you identify threat actors most likely to target the organization. This helps the pentester better prioritize vulnerabilities found during the engagement.

Compare bug bounty programs and a pentest

Example answer:

Bug bounty programs can typically find more vulnerabilities over time than a pentest because they involve continuous testing. You will also get a more diverse group of skill sets, and the payouts of many bug bounty programs are far less than the cost of a single pentest.

What is an HTTP desync attack?

Example answer:

HTTP desync attacks abuse the method by which a chain of HTTP servers interprets consecutive requests, especially around the boundaries of requests. As an example, an attacker could send a request with a transfer-encoding header that doesn't have the values specified in **Request for Comments (RFC)** *7230*. This can help the attacker hide the encoding of their payload from the WAF.

What is the difference between vertical and horizontal privilege escalation?

Example answer:

Horizontal privilege escalation refers to bypassing the authentication mechanism for users who have the same level of privilege and taking over their accounts.

Vertical privilege escalation refers to escalating privilege to a higher level of access, such as a standard user now having the same level of access as the administrator account.

How often should organizations have an external pentest performed?

Example answer:

This answer depends on their compliance requirements, but generally, this should happen at least once a year and preferably on a quarterly basis. One thing you will notice when you're working as a pentester is that many companies will not fix any of the issues you report, so you might come back a year later and identify the same issues.

What are the legal considerations for pentests?

Example answer:

With pentests, you need to have a contract in place before starting the engagement. The contract is often referred to as your *get-out-of-jail-free card*, but keep in mind that you could still be arrested for performing a pentest even if it's authorized, especially if your actions go away from the ROE.

Some other key legal considerations are outlined here:

- Does the client really own the systems and/or applications they want you to test?
- Will the client assume liability for any interruptions or damage that occur as a result of the pentest, or are you responsible?
- What happens when third-party data or services are damaged as a result of the pentest? Who is responsible?
- Do you need a private investigator license to perform a pentest in that geographic location?
- Which jurisdiction will be recognized for the pentest? For example, if you are testing offices in Alabama and Virginia, which state's laws will apply to the engagement?
- Who owns any new methods or tools that are developed as a result of the pentest engagement?
- Is there a duty to warn third parties about pentest results based on the findings? For example, you discover a high-severity zero-day exploit as a result of a pentest. Do you report it?

What is a buffer overflow attack?

Example answer:

Buffers are memory storage regions that temporarily hold data while it is being transferred from one location to another. A buffer overflow occurs when the volume of data exceeds the storage capacity of the memory buffer. As a result, the program attempting to write the data to the buffer overwrites adjacent memory locations.

For example, a buffer for login credentials may be designed to expect username and password inputs of 8 bytes, so if a transaction involves an input of 10 bytes (that is, 2 bytes more than expected), the program may write the excess data past the buffer boundary.

As you can see, the questions you might be asked during an interview for pentester roles can vary, but the main thing to keep in mind is that for more junior-level roles, the interview is typically focused on knowledge with a small hands-on component. For more senior-level interviews, you can expect a more hands-on interview.

Summary

In this chapter, you learned what a pentester is, the range of salaries in the US for pentesting, and common questions you might be asked during an interview. It's important to remember that the questions listed in this chapter cover entry-level through principal pentester roles, so you might not be asked all the questions from this chapter during your job interview.

In the next chapter, you will learn about digital forensic analyst careers.

Join us on Discord!

Read this book alongside other users. Ask questions, provide solutions to other readers, and much more.

Scan the QR code or visit the link to join the community.

`https://packt.link/SecNet`

5

Digital Forensic Analyst

In this chapter, you will learn what a digital forensic analyst is and the average salary range for this career in the United States. You will also learn about career progression options and learn common interview questions for the role.

The following topics will be covered in this chapter:

- What is a digital forensic analyst?
- How much can you make in this career?
- What other careers can you pursue?
- Common interview questions for a digital forensic analyst career

What is a Digital Forensic Analyst?

Digital forensic analysts are tasked with collecting, preserving, and analyzing digital evidence. They might work with **incident response teams** (**IRTs**) to investigate incidents and attempt to identify threat actors responsible for an attack. Digital forensic analysts may also work with law enforcement agencies to help in criminal investigations, including crimes against children, and help companies in civil and administrative investigations. They may also be hired by law firms to conduct **electronic discovery** (**e-discovery**) work, where the analyst collects electronic evidence to be used in civil cases. An example of this might be collecting evidence on the financial activity of one spouse during a divorce proceeding to help the attorneys prove how much money they have in their accounts. As a digital forensic analyst, you will analyze operating systems such as Windows, macOS, Linux, and mobile OSs, analyze volatile and non-volatile data, and work with forensic tools such as EnCase (`https://www.opentext.com/products/encase-forensic`) and Autopsy (`https://www.sleuthkit.org/autopsy/`).

The **Digital Forensics and Incident Response (DFIR)** *Diva* blog (`https://dfirdiva.com/`) contains listings of free and low-cost resources for you to gain hands-on experience in conducting forensic investigations.

How much can you make in this career?

The salary range for a digital forensic analyst in the US depends on several factors, such as your location, the size of the company you work for, the certifications you hold, college degrees, and your skills. In the US, the average salary to expect is between $74,300 and $122,000. If you work in an IRT and have some experience, your base salary in the US may be in the six-figure range. Typically, you would need several years of experience to be qualified for a dedicated forensic position, but you can also do forensic tasks as part of your work on an incident response team, so starting with an SOC Analyst position could be the best choice for you.

What other careers can you pursue?

The skills learned as a digital forensic analyst can prepare you for other careers such as Penetration Tester, Malware Analyst, Cybersecurity Manager, and senior executive roles as you advance in your career, such as **Chief Information Security Officer (CISO)**. As a digital forensic analyst, you can also use your skills to work on an IRT.

Common interview questions for a digital forensic analyst career

In this section, you will learn some of the most common interview questions that are posed in relation to digital forensic analyst jobs. We will present a list of these here.

What is the chain of custody (CoC)?

The chain of custody in digital forensics refers to the documented process that records the sequence of custody, control, transfer, analysis, and disposition of digital evidence. It ensures that the evidence is collected, preserved, and handled in a manner that maintains its integrity and authenticity.

This documentation includes details such as who collected the evidence, when it was collected, how it was transported, who accessed it, and any changes made to it. Maintaining a strict chain of custody is important for the admissibility of evidence in legal proceedings, as it helps to prove the evidence has not been tampered with or contaminated.

You can view an example of a chain of custody form at this link: `https://www.nist.gov/document/sample-chain-custody-formdocx`.

Which tools can be used to recover deleted files?

These are some tools that can be used to recover deleted files.

- EaseUS (`https://www.easeus.com/`)
- Advanced Recovery
- Disk Drill (`https://www.cleverfiles.com/data-recovery-software.html`)
- Recoverit (`https://recoverit.wondershare.com/`)
- Recuva (`https://www.ccleaner.com/recuva`)

Can you provide examples of some common hashing algorithms?

The SHA-2 family, particularly SHA-256 and SHA-512, is currently considered secure and is commonly used in various applications such as SSL/TLS certificates, digital signatures, and blockchain technology.

What is data carving?

Data carving is conducted in a forensic investigation to identify deleted information on a system. Some forensic tools offer data-carving capabilities, and these typically identify file headers and footers to recover intact files, meaning the files have been deleted but not overwritten yet by new data. As a forensic investigator, you can also conduct manual data carving, whereby you pull fragments from previous files in slack space.

What is data mining?

Data mining is just the process of pulling out specific information from large datasets. In digital forensics, mining can be used to collect correlating data on a suspect. As an example, let's suppose a suspect visits an internet café to use its computers to commit criminal activity. Let's pretend the internet café doesn't have user accounts, so the only way to identify the person responsible for the crimes is to collect information from the computer, including timestamps, and then correlate that information with security cameras in the area to narrow down the suspect list and identify the person committing the illegal activity.

What are some considerations around forensic investigations in the cloud?

Considerations include jurisdiction, **Cloud Service Providers (CSPs)**, and multi-tenancy. CSPs host data all over the world, which causes jurisdictional issues. For example, you might work with law enforcement in the US and have a warrant to seize data on a suspect, but the data might be stored in Russia, which doesn't recognize your warrant. Another consideration is the CSPs themselves, as they control the hardware and the logging capability in **platform-as-a-service** (**PaaS**) and **software-as-a-service** (**SaaS**) deployments. CSPs might also sanitize log files from customers and have policies restricting access to log files. Multi-tenancy is another challenge because other organizations do not want you as an investigator to accidentally access their data from the cloud.

Can you name some common encryption algorithms that are used to encrypt data?

Common encryption algorithms used to encrypt data include **Advanced Encryption Standard (AES)**, **Rivest-Shamir-Adleman (RSA)**, and **Data Encryption Standard (DES)**. AES is a symmetric key encryption algorithm that supports key sizes of 128, 192, and 256 bits and is widely used, due to its high security and efficiency.

RSA is an asymmetric encryption algorithm that uses a pair of keys (public and private) and is commonly used for secure data transmission and digital signatures. DES, though once standard, uses a 56-bit key and has largely been replaced by more secure algorithms like AES and (**Triple DES (3DES**), due to vulnerabilities that make it susceptible to brute-force attacks.

Explain RSA to me and how it differs from AES

RSA is an asymmetric encryption algorithm that uses a pair of keys: a public key for encryption and a private key for decryption. This key pair enables secure data transmission and digital signatures, as data encrypted with the public key can only be decrypted with the corresponding private key.

AES is a symmetric encryption algorithm that uses the same key for both encryption and decryption, making it faster and more efficient to encrypt large amounts of data. The primary difference between the two lies in their key management: RSA's asymmetric nature makes it suitable for secure key exchange and authentication, while AES's symmetric nature makes it ideal for high-speed data encryption and decryption.

What is SIFT?

The **Sans Investigative Forensic Toolkit (SIFT)** is a forensic workstation from SANS that comes with a number of pre-installed forensic and IR tools. Using pre-made images such as SIFT and the Volatility framework (`https://github.com/volatilityfoundation/volatility`) can save you time in setting up your forensic investigation lab.

What is timeline analysis?

Timeline analysis is a sequence of events on a system or group of systems that allows the investigator to see what happened and when, along with which events happened just before or after an incident.

What is metadata?

Metadata is commonly known as data about data. There are three types of metadata, which are **descriptive**, **administrative**, and **structural**. Descriptive metadata contains information about a file, such as the file author, keywords, and title. Administrative metadata contains the ownership and rights management of a file and which program was used to create the file. Structural metadata contains relational information on file data. In digital forensics, metadata can be used to identify the security settings of a file, and in the case of an email thread, metadata can be used to track the email origin and which other systems the email has passed through to its destination.

You respond to a ransomware attack incident. Walk me through your process to preserv, collect, and analyze evidence in this situation, and be sure to explain how you ensure the chain of custody.

In responding to a ransomware attack, the first step is to preserve the current state of all affected systems to prevent further damage and loss of evidence. This involves isolating infected machines from the network to stop the spread of ransomware. Next, I would create forensic images of the affected systems, capturing the state of the hard drives and memory at the time of the attack.

To collect evidence, I also focus on gathering system logs, network traffic data, and any ransom notes or related communications from the threat actors. I would use forensic tools like **EnCase**, **FTK Imager**, and **Volatility** for this purpose. The evidence collected could include details of the ransomware's behavior, such as the files encrypted, processes initiated, and any communications with **command-and-control (C2)** servers.

During the analysis phase, I attempt to identify the ransomware variant by examining file signatures and ransom notes. I analyze logs and network traffic to determine the initial infection vector and track the ransomware's activity within the network. I also look for any artifacts that can provide clues about the attacker's identity or methods.

To maintain the chain of custody, I meticulously document every step taken during the investigation. This includes recording the date, time, and personnel involved in each stage of evidence handling. Each piece of evidence is labeled with a unique identifier, and detailed logs are kept of who accessed the evidence, when it was accessed, and any actions taken with it. This documentation ensures that the integrity and authenticity of the evidence are preserved, making it admissible in legal proceedings.

How do you get indicators of compromise (IOCs) from analyzing malware samples?

You can get IOCs by using static analysis. The first step should be to obtain a hash of the malware file and then search online databases, such as VirusTotal, to see whether anyone else has already done a write-up on the malware sample. This step can save you hours of frustration in your investigation. You can then use a tool such as Sysinternals (`https://docs.microsoft.com/en-us/sysinternals/`), coupled with **regex**, to analyze the strings of the malware sample to look for **Internet Protocol** (**IP**) addresses, suspicious **Uniform Resource Locators** (**URLs**), and file paths. If you are not familiar with using the Sysinternals suite, this YouTube video from Mark Russinovich provides an overview of using Sysinternals for malware analysis: `https://www.youtube.com/watch?v=vW8eAqZyWeo`.

What is the difference between static and dynamic malware analysis?

Static malware analysis is used to analyze the malware sample and its code without executing it. The forensic investigator might be limited in seeing what capabilities the malware has with simple static analysis, so dynamic malware analysis is used to analyze the behavior of the malware sample. More advanced static analysis can be used to dissect the malware down to assembly language, but this reverse engineering is time-consuming and not pragmatic for many investigations that are part of IR.

What is a PE file?

Portable executable (**PE**) is the standard Windows file format for executable files, **dynamic-link libraries** (**DLLs**), and object code for both 32-bit and 64-bit Windows operating systems.

How would a piece of malware maintain persistence?

Common methods of maintaining persistence include:

- **Registry keys**: Modifying or creating Windows Registry keys, like **Run** or **RunOnce**, to execute automatically during the system's startup process.

- **Scheduled tasks**: Creating or modifying scheduled tasks that trigger it at specified intervals or during system events.

- **Services**: The malware can install itself as a legitimate system service, which ensures it runs with elevated privileges and starts with the system.

- **Startup folders**: Malware can place executable files or scripts in startup folders that run automatically upon login.

- **Bootkits**: By infecting the boot sector, the malware can start before the operating system loads, making it difficult to detect and remove with traditional antivirus solutions.

- **DLL hijacking**: The malware can exploit the way applications load DLLs, replacing legitimate DLLs with malicious ones.

- **Persistence in firmware**: More advanced malware can reside in firmware components like the BIOS or UEFI, making it resistant to hard drive wipes and OS reinstallation.

Can you name some items you would carry in your forensic response kit?

The contents of a forensic response kit will vary based on the investigator, but some items you will want to include are antistatic bags, your forensic laptop, dongles, a screwdriver toolkit, extra cables, Faraday bags or commercial aluminum foil, write blocking devices, storage media, gloves, a digital camera to record the condition of the scene when you arrive, a notepad, and evidence paperwork, such as labels or tags and CoC forms.

What are the two main types of data you deal with as a digital forensic investigator?

The two main types of data are volatile and non-volatile.

What is volatile data?

Volatile data is temporary data on your digital device that is dependent upon having a steady power supply. If the power is interrupted at all, this data can be lost.

Some examples of volatile data include the system time, a listing of users logged onto the system, a list of files that are open, information on the network, information on processes running on the system, process-to-port mapping, services running, a list of drivers on the system, a history of the command run on the system, and the contents of the clipboard.

What is non-volatile data?

Non-volatile data is data that will remain on a system, even if the power supply is interrupted. This type of data can be stored on secondary storage devices, such as memory cards and a hard disk. Examples of non-volatile data include slack space, hidden files, swap files, the `index.dat` file, clusters that are unallocated, partitions that are not being used, your registry settings, and system event logs.

Can you provide examples of artifacts you can get from analyzing RAM?

Some artifact information you can get from analyzing RAM includes encryption keys, passwords, IP addresses, browsing history, cleartext data, configuration information, and commands that were entered.

In which situations can duplicate evidence suffice as evidence?

Situations in which duplicate evidence will be accepted include if the original evidence is destroyed due to a fire, flood, or other disaster in the normal course of business. Duplicate evidence can also be used if the original evidence is in the possession of a third party.

Explain the difference between civil, criminal, and administrative cases in digital forensic investigation, and explain what you would do in each one as a digital forensic investigator.

- **Civil cases**: Civil cases typically involve disputes between individuals or organizations over rights, obligations, or damages. These can include cases like Intellectual Property theft, breach of contract, or employee misconduct. In a civil case, my role would be to collect and analyze digital evidence to support the legal arguments of one party. This involves preserving electronic data, performing forensic imaging of devices, analyzing emails, documents, and logs, and providing expert testimony if required. I must ensure that the evidence is handled meticulously to maintain its integrity for court admissibility (e.g., Chain of custody).

- **Criminal cases**: Criminal cases involve actions that are considered offenses against the state or public, such as hacking, fraud, child exploitation, or cyberstalking. In a criminal case, my primary focus is on uncovering and analyzing evidence to help establish whether a crime has occurred and to help identify the perpetrator. This includes securing and examining digital devices, network logs, and digital communications. I work closely with law enforcement agencies, follow strict protocols to maintain the chain of custody, and may testify in court regarding the findings and methodologies used during the investigation.

- **Administrative cases**: Administrative cases involve internal investigations within organizations or regulatory compliance issues. These might include policy violations, internal fraud, or compliance with data protection regulations. In an administrative case, my duties involve conducting internal investigations to uncover policy breaches or regulatory violations. This includes analyzing user activities, examining logs and emails, and identifying unauthorized actions. I document the findings and provide recommendations to prevent future incidents. In some cases, I may need to collaborate with regulatory bodies or legal teams within an organization.

Which amendment in the US protects against illegal search and seizure by government authorities?

The Fourth Amendment to the United States Constitution protects against illegal search and seizure by government authorities. This amendment ensures that government authorities, including law enforcement, must obtain a warrant based on probable cause before conducting searches and seizures. The warrant must be specific, detailing the exact location to be searched and the items or persons to be seized. The Fourth Amendment's goal is to protect individuals' privacy and property from arbitrary or unjustified intrusions by the government, thereby upholding citizens' rights to privacy and security. Violations of this amendment can result in evidence being excluded from court proceedings under the exclusionary rule, which is designed to deter illegal conduct by law enforcement.

What is the primary purpose of the first responder?

The main goal of the first responder is to secure a scene until investigators arrive. This is done to help protect evidence from contamination or theft.

Can you provide examples of some forensic tools?

- **EnCase** is used for data acquisition, analysis, and reporting. EnCase is one of the most popular digital forensic tools.

- **FTK (Forensic Toolkit)** is a tool used to analyze and index digital evidence. It supports data carving, email analysis, and decryption.

- **Autopsy** is an open-source digital forensics platform that provides a **Graphical User Interface (GUI)** to **The Sleuth Kit (TSK)**. It is used to analyze disk images and recover deleted files and partitions in forensic investigations.

- **Volatility Framework** is an open-source tool for memory forensics that allows investigators to analyze RAM dumps to detect malware.

- **X-Ways Forensics** is used for disk imaging, data recovery, and analysis. It is known for its speed in handling large datasets.

- **Wireshark** is useful for network forensics to identify suspicious activities in network traffic.

- **Cellebrite UFED** is used for mobile device forensics and supports data extraction, decoding, and analysis from a wide range of mobile devices.

What is FTK Imager?

FTK Imager is a popular forensic tool that helps an investigator acquire and analyze the files and folders found on system hard drives, network drives, and **Compact Disk—Read-Only Memory. (CD-ROM)/Digital Versatile Disc (DVD)**. The tool also helps investigators analyze forensic images and memory dumps. Some other capabilities of FTK Imager include the ability to create hashes of files, recover and review deleted files from the Recycle Bin in the Microsoft Windows operating system, and export files and folders from captured forensic images to disk.

What is EnCase?

EnCase is a multi-purpose digital forensic platform that includes many useful tools to support your digital forensic investigation.

Which law in the US deals with fraud and related activity in connection with computers?

Title 18 US Code subsection 1030—or, more appropriately written, *18 USC §1030*—deals with fraud and other activity in connection with computers. This is also called the **Computer Fraud and Abuse Act (CFAA)**.

Which federal law in the US covers Child Sexual Abuse Material (CSAM)?

In digital forensic investigations, **CSAM** refers to any content that depicts sexually explicit activities involving a child. This type of material is illegal to possess, distribute, or produce and is a significant focus of many digital forensic investigations. Title 18 US Code subsection 2252A (18 USC §2252A) covers CSAM.

If an investigator needs to obtain information from a service provider (SP), such as billing records and subscriber information from a victim's computer, what type of warrant is issued?

An **SP search warrant** allows the investigator or first responder to obtain victim information such as billing records and subscriber information.

What is a platter?

Platters are circular metal disks that are mounted in the drive enclosure.

What are sectors?

Sectors are small, physical storage units located on the hard disk platter; they are 512 bytes long.

What is slack space?

When a filesystem allocates an entire cluster for a file, but the file size is much smaller than the full cluster available, the remaining area is known as slack space.

What is a GUID?

A **globally unique identifier** (GUID) is a 128-bit unique number generated by Windows that is used to identify things such as **Component Object Model** (COM) DLLs, primary key values, browser sessions, and usernames. A GUID is sometimes known as a **universally unique ID** (UUID).

What is file carving?

File carving is a technique used to recover files and fragments of files from an unallocated portion of the hard disk. This technique can be used if you can't find any file metadata.

What type of image file format is lossless?

Portable Network Graphics (PNG) is a lossless image format that was intended to replace the **Graphics Interchange Format** (GIF) and **Tagged Image File Format** (TIFF) formats.

What type of image file starts with a hexadecimal (hex) value of FF D8 FF?

Joint Photographic Experts Group (JPEG) files start with this hex format.

What is the Master Boot Record (MBR)?

The **MBR** holds information about partitions, the bootloader code, and information on filesystems.

Can you explain the differences in the boot process between macOS, Windows, and Linux?

- macOS boot process:

    1. **Firmware initialization**: The process begins with the system firmware (**EFI (Extensible Firmware Interface)**). On modern Macs, this is the **Unified Extensible Firmware Interface (UEFI)**. The firmware performs hardware checks and initializes the hardware components.

    2. **Boot Manager**: The firmware loads the Boot Manager, which displays a list of available boot devices (holding the Option key at startup allows you to choose a different boot disk).

    3. **Boot loader**: The Boot Manager locates and loads the boot.efi file from the selected startup disk. boot.efi initializes and loads the kernel.

    4. **Kernel initialization**: The macOS kernel is loaded into memory. The kernel initializes the hardware, mounts the root filesystem, and starts the launchd process (which replaces the older init process).

    5. **User space initialization**: launchd starts system and user-level services and applications, completing the boot process.

- Windows boot process:

    1. **Firmware initialization**: The process begins with the BIOS or UEFI firmware performing the **Power-On Self Test (POST)** to check hardware components.

    2. **Boot manager**: For BIOS systems, the **Master Boot Record (MBR)** loads the Boot Manager. For UEFI systems, the firmware loads the Windows Boot Manager (bootmgfw.efi).

    3. **Boot loader**: The Windows Boot Manager reads the **Boot Configuration Data (BCD)** to determine the boot options. It loads the `winload.exe` (or `winload.efi` for UEFI).

4. **Kernel initialization**: winload.exe loads the Windows kernel (`ntoskrnl.exe`), the **Hardware Abstraction Layer** (**HAL**), and essential drivers. The kernel initializes and mounts the Windows Registry.

5. **Session Manager**: The kernel starts the Session Manager Subsystem (smss.exe), which initializes system sessions. smss.exe starts the Winlogon process (winlogon.exe), which handles user login.

6. **User space initialization**: `winlogon.exe` starts the Service Control Manager (`services.exe`), which loads and starts services. The user logs in, and the Windows Explorer shell starts, completing the boot process.

- **Linux boot process**:

    1. **Firmware initialization**: The process starts with the BIOS or UEFI firmware performing POST to check hardware components.

    2. **Boot loader**: The firmware loads the boot loader (GRUB is a common choice). GRUB displays the boot menu and loads the selected Linux kernel and initial RAM disk (`initrd` or `initramfs`).

    3. **Kernel initialization**: The Linux kernel is loaded into memory and initializes hardware. The kernel mounts the initial RAM disk, which contains essential drivers and tools needed for the next stage.

    4. **Initial RAM disk**: The initial RAM disk sets up a temporary root filesystem and prepares the environment for the real root filesystem. It loads the necessary drivers and initializes devices needed to mount the real root filesystem.

    5. **Init system**: The kernel hands over control to the init system (`systemd`, `Upstart`, or `SysV init`, depending on the distribution). The init system starts system services and user-space processes according to predefined scripts or configurations.

    6. **User space initialization**: The init system manages system and user services, networking, and other tasks. It eventually starts the **graphical user interface** (**GUI**) or **command-line interface** (**CLI**), completing the boot process.

Which National Institute of Standards and Technology (NIST) document covers sanitation techniques for the media?

NIST **Special Publication** (**SP**) *800-88* guidelines cover proper techniques to sanitize media.

What is live data acquisition?

This is the process of acquiring volatile data (for example, RAM) from a computer that is turned on, but that is either locked or in sleep mode. You conduct live data acquisition (sometimes called live box forensics) to acquire volatile data because it is lost when a system suffers a power outage or when a user turns the system off.

Why would you use write protection when acquiring evidence?

Write protection is used when acquiring images so that you do not alter the original data.

Can you name some functions that are offered with dcfldd and not dd?

Some of the functions offered are status output, hashing on the fly, flexible disk wipes, verifying that a target drive is a bit-for-bit match, outputting to multiple disks or files at the same time, splitting output into multiple files with additional configurability, and piping the output.

Can you name some anti-forensic techniques?

Some common anti-forensic techniques include using encryption, data cleaning, using packers such as **Ultimate Packer for eXecutables** (**UPX**), using **The Onion Router** (**TOR**), altering time-stamps, and using steganography.

UPX is used to compress files. A threat actor might use a packer such as this to try to get their malware past security tools that scan for specific file signatures.

The TOR browser is used to anonymize the internet activity of a user. This can be helpful for journalists in oppressive regimes. A threat actor might use TOR during their attack to obfuscate their actual IP address.

Where is the Google Chrome history file located?

The history file in Chrome is found in `%USERS%/AppData/Local/Google/Chrome/User Data/`.

What information can you gather from the Chrome history file?

Some information that can be obtained from the history file includes any typed URLs, keyword searching, and downloads of files.

What does the HKEY_CLASSES_ROOT registry key contain?

This contains the file extension association information and **programmatic ID** (**ProgID**), **class ID** (**CLSID**), and **interface ID** (**IID**) data.

What does the HKEY_CURRENT_USER registry contain?

This registry key contains the configuration information (that is, wallpaper preference, screen colors, and display settings) related to the user currently logged on.

What does the HKEY_USERS registry key contain?

This registry key contains information about all the active user profiles on the system.

Which tools can be used to analyze the Windows Registry?

Some tools you could use are **regedit**, **RegRipper**, **Process Monitor (ProcMon)**, **Registry Viewer**, and **ProDiscover**.

Can you provide a registry key that threat actors frequently add malicious entries to in order to maintain persistence?

The *Run* key often has malicious entries. For example, **APT29** commonly adds a **spool.exe** entry to the registry key.

What sort of data should you collect in the event of a website attack?

Some data that should be collected includes the following:

- The date and time at which **HyperText Transfer Protocol (HTTP)** requests were sent
- The source IP address of the request
- Which HTTP method was used (**GET**, **POST**, and so on)
- HTTP query information
- A full set of HTTP headers and the full HTTP request body
- Any event logs
- Any file listings and timestamps

What is database forensics?

Database forensics is the examination of databases and related metadata using forensically sound practices to ensure the findings are admissible in a court of law.

Where does the Microsoft SQL (MSSQL) server store data and logs?

It stores them in primary data files (**Main Database Files (MDFs)**), secondary data files (known as **SQL, Server Secondary Database Files**, or **NDFs**), and transaction log data files (**Log Database Files (LDFs)**).

In which directory does the MySQL Server store status and log files?

It stores status and log files, along with other data managed by the server, under the **data** directory.

Can you list the categories of cloud crimes?

The categories of cloud crimes are listed here:

- **Cloud as a subject**, which refers to a crime in which attackers try to compromise the security of a cloud environment to steal data or inject malware—for example, stealing credentials of a cloud account and leveraging the credentials to delete or modify data stored in the cloud environment.
- **Cloud as an object**, which refers to an attacker leveraging the cloud environment to conduct an attack against CSPs. **Distributed Denial of Service (DDoS)** attacks are an example of an attack that is leveraged against CSPs.
- **Cloud as a tool**, which refers to when an attacker uses one compromised cloud account to attack other accounts. In these situations, the source and destinations of the attack can yield evidence for your case.

What are some common cloud threats?

Some common threats involved in using cloud environments include data breaches, data loss, abuse of native cloud services for attacks, insecure **application programming interfaces (APIs)**, security misconfigurations, a lack of accountability to keep data safe, not clearly identifying who owns the responsibility for the security of data, a lack of user ID federation, a lack of visibility, multi-tenancy security concerns, and a lack of compliance.

Can you provide examples of crimes that are supported by email capabilities?

Some crimes include **business email compromise (BEC)**, identity theft, cyberstalking, and crimes targeting children.

Can you name some tools used to collect and analyze emails?

Tools that can be used for email collection and analysis include Stellar Phoenix Deleted Email Recovery, FTK, Paraben E-mail Examiner, and Kernel for Outlook PST Recovery.

What is the purpose of the Stellar Phoenix Deleted Email Recovery software?

It helps you recover lost or deleted emails from MS Outlook data (**Personal Storage Table (PST)**) files and **Outlook Express Mail Database (DBX)** files.

Paraben's E-mail Examiner helps you examine different email formats, including Outlook (PST and **Offline Storage Table (OST)**), Thunderbird, Outlook Express, Windows Mail, and more. The tool allows you to analyze message headers, bodies, and attachments. It also helps recover emails from deleted folders, offers support for advanced searching and reporting, and offers an export capability to PST and other formats.

What information can you find on a subscriber identity module (SIM) card?

You can locate information such as contacts, messages, timestamps, an **integrated circuit card ID (ICCID)**, the last numbers dialed, and the SP name.

What is the International Mobile Equipment Identifier (IMEI)?

The IMEI is a 15-digit unique number on the handset that identifies mobile equipment.

What are common sources of digital forensic data on mobile devices?

Some locations include:

- Call logs and contact information
- Messages and communication apps
- Emails and email accounts
- Multimedia files (photo, video, and audio)
- Geolocation data
- Browsing history
- Data from installed apps
- System logs
- Configuration files
- File storage and cloud storage
- Social media accounts
- Memory
- SIM card

Can you name a tool that can be used to gather information from Facebook and Twitter?

Bulk Extractor is a tool that can be used to collect and analyze social media artifacts from a captured memory file or forensic image.

What are some benefits of solid-state drives (SSDs)?

In comparison to traditional hard drives, SSDs offer increased reliability, weigh less, increase the data access speed, and help reduce power consumption.

What is the master file table (MFT)?

The MFT tracks the files in the volume and essentially manages them.

You are asked to investigate a potential insider threat where an employee is suspected of stealing sensitive data. How would you approach this investigation?

I would begin by securing and imaging the employee's devices to preserve evidence. Next, I would analyze network logs, email records, and file access logs to identify any suspicious activities. I would also review the employee's recent activity for any large file transfers or unusual access patterns. I would also interview relevant personnel and compile a detailed report with my findings. Throughout this process, I would ensure evidence preservation and the chain of custody.

A company server has been compromised and used to distribute malware. What steps would you take to investigate this incident?

I would start by isolating the compromised server. Then, I would create a forensic image of the server for analysis. I would examine system logs, network traffic, and installed software to identify the malware's origin and behavior. Additionally, I would look for any signs of lateral movement or data exfiltration. I would ensure the chain of custody and document my findings in a report that also includes recommendations for mitigation.

You need to recover deleted files from a suspect's hard drive. What tools and techniques could you use for this task?

Digital forensic tools like **EnCase**, **FTK Imager**, or **Autopsy** could be used to create an image of the hard drive and perform file recovery. These tools can help locate and reconstruct deleted files by examining the filesystem and unallocated space. I would also analyze file metadata and timestamps to understand the context and relevance of the recovered files. Throughout the process, I would ensure the chain of custody and report on my findings.

You are tasked with analyzing a mobile device for evidence in a fraud case. What steps would you take to preserve and analyze the data on the mobile device?

I would place the mobile device in airplane mode to prevent remote tampering. Then, I would use forensic tools like **Cellebrite** or **Oxygen Forensic Suite** to create a forensic image of the device. I would analyze the image for relevant data, such as text messages, call logs, emails, and app data. I would also examine the device's location history and installed applications for additional evidence. I would document all findings, ensure the chain of custody, and report on my findings.

You are asked to investigate a potential intellectual property theft by a former employee. How would you approach this investigation?

I would start by examining the former employee's computer and network activity during their final weeks of employment. This includes reviewing file access logs, email records, and any external storage devices used. I would also analyze any data transfers to cloud services or personal email accounts. Interviews with current employees and reviewing exit procedures would provide additional context. I would ensure the chain of custody and report my findings.

You need to perform a forensic analysis on a virtual machine (VM) suspected of being used in an attack. What steps would you take to analyze the VM?

I would begin by creating a snapshot of the VM to preserve its current state. Then, I would use forensic tools like **Volatility** or **FTK Imager** to create an image of the VM's disk and memory. I would analyze the disk image for malware, suspicious files, and log files. Memory analysis would help identify running processes, network connections, and any injected code. I would ensure the chain of custody throughout my investigation and report on my findings.

You are investigating a network intrusion and find a suspicious executable file. How would you analyze this file to determine its purpose and impact?

I would start by creating a hash of the file and searching for it in malware databases like **Virus Total** (https://www.virustotal.com/gui/home/upload). If the file is not known, I would perform static analysis using tools like strings, **PEiD**, or **IDA Pro** to examine its structure and embedded data. Next, I would conduct dynamic analysis in a controlled environment to observe the file's behavior, such as network connections, file modifications, and registry changes. I would document my findings and use them to understand the intrusion's scope and impact.

Summary

In this chapter, you learned about the job of a digital forensic analyst and what to expect in terms of an average salary. You also learned the two main types of data that can be collected in a digital forensic investigation—volatile and non-volatile data—and some of the most common interview questions you may face from a hiring manager. You also learned about some tasks and investigations you might be involved in as a digital forensic investigator.

In the next chapter, we will discuss a cryptographer/cryptanalyst's career path.

Join us on Discord!

Read this book alongside other users. Ask questions, provide solutions to other readers, and much more.

Scan the QR code or visit the link to join the community.

https://packt.link/SecNet

6

Cryptographer/Cryptanalyst

This chapter will cover two job roles, cryptographer and cryptanalyst, blended under the title cryptographer since most open private sector job postings list *cryptographer* as the job title in demand. As a cryptographer, you will also be doing cryptanalyst work in attempting to break encryption. In this chapter, you will learn about what cryptographers do, where they might work, and the average salary range for cryptographers in the US. You will also learn about the career progression options and common interview questions for the role.

The following topics will be covered in this chapter:

- What is a cryptographer?
- How much can you make in this career?
- What other careers can you pursue?
- Common interview questions for cryptographers

What is a Cryptographer?

Cryptographers write and crack the encryption code used to protect data. In a cryptographer role, you will help to develop better algorithms to help protect data from threats. Depending on the organization you work with, your day-to-day work will vary, but you will be helping to protect the confidentiality of data. Cryptographers also help to develop mathematical and statistical models that can help organizations locate and disrupt threats to their systems.

Some of your day-to-day work as a cryptographer will include:

- Identifying weaknesses in existing cryptography systems and identifying ways to better secure them

- Conducting testing for cryptology theories

- Improving data security across the organization

- Deploying symmetric and asymmetric cryptography

- Managing the organization's encryption implementation, especially as it relates to code and third-party products

- Conducting the training of other departments to help them implement better encryption practices

- Developing new encryption solutions

To be successful in a cryptography career, you should have hands-on experience in operating systems and computer networking and know at least one programming language. You will also need to be familiar with different encryption algorithms, hashing, number theory, key exchange, data structures, and digital signatures, and have strong mathematical skills in areas such as linear algebra. In addition to these hard skills, you will need soft skills such as the ability to work well in teams, effective communication with different stakeholders, problem-solving and critical thinking skills, and good time management skills.

Full-time cryptography roles are typically found in public sector (government) work. In the US, cryptographers might work for government entities such as the **National Security Agency (NSA)**, the **National Institute of Standards and Technology (NIST)**, or the **Department of Defense (DoD)**.

Certifications available for this career path include **Certified Blockchain Security Professional (CBSP)** and **EC-Council Certified Encryption Specialist (ECES)**.

If you are looking to gain hands-on experience working through cryptography challenges, the Cryptopals website has free challenges around cryptography (`https://cryptopals.com/`).

SimpliLearn also has free cryptography training available on their YouTube channel at this link: `https://youtu.be/C7vmouDOJYM`.

How much can you make in this career?

The average salary for cryptographers and cryptanalysts in the US can vary based on the organization that you work for, your location, years of experience, certifications, college degrees, and more. The salary in the US varies between $56,400 and $167,500. With the right skills and experience, some people are making over $200,000 in these careers.

What other careers can you pursue?

A career as a cryptographer can help you advance into other cybersecurity careers such as a Penetration Tester, an incident responder, and a malware reverse engineer. This career is also a natural progression into a crypto and blockchain security researcher career, where you may be researching emerging technology and/or threats, contributing to research publications, and developing new approaches to managing threats. It can also help you move into a career as a blockchain developer.

Common interview questions for cryptographers

The following questions are designed to assess your fundamental knowledge of cryptography. In job interviews, you may also be asked to solve cryptography challenges. The hands-on assessment will depend upon the employer and the role you are applying for.

What is the difference between cryptography, cryptology, and cryptanalysis?

Cryptology is a broader field that encompasses both cryptography and cryptanalysis. It is the science of studying and analyzing cryptographic systems and protocols. Cryptology covers the design, implementation, and analysis of secure communication methods and includes both the creation of cryptographic algorithms and the breaking of these algorithms to find vulnerabilities.

Cryptography is the practice and study of techniques for securing communication and data from threat actors. It involves creating algorithms, protocols, and systems that ensure confidentiality, integrity, authentication, and non-repudiation. Cryptography includes various methods, including symmetric key cryptography (e.g., AES), public key cryptography (e.g., RSA), and hashing functions (e.g., SHA-256).

Cryptanalysis is the art and science of analyzing cryptographic systems to uncover weaknesses or to break cryptographic algorithms. It involves studying the underlying mathematics and algorithms used in cryptography to find vulnerabilities, perform attacks, and decrypt information without prior knowledge of the secret key. Cryptanalysis includes various techniques, such as brute force attacks, differential cryptanalysis, linear cryptanalysis, and side-channel attacks.

To sum these up, cryptography focuses on creating secure systems, cryptanalysis on breaking them, and cryptology is the overarching field that includes both the creation and breaking of cryptographic systems.

What is the difference between encoding, hashing, and encryption?

Encoding is just the process of converting data from one format to another. **ASCII**, **Base64**, and **Unicode** are examples of encoding algorithms. In encoding, the same algorithm is used to encode and decode the data, which means an attacker would just need to have the data sample to be able to decode it.

Encryption is the process of using a cryptographic key to scramble your data so it is unreadable. Symmetric encryption involves using a shared key where both the sender and receiver know the key. The shared key is then used to encrypt and decrypt the data. Asymmetric encryption uses a public and private key pair. The public key is known (hence the name *public*) and is used to encrypt the data, and the receiver's private key is used to decrypt the data. **Rivest-Shamir-Adleman (RSA)** is a well-known asymmetric encryption algorithm.

Hashing is a one-way function in which a string of information is run through a hashing algorithm, and it produces a fixed-length output based on the algorithm you are using and the input data. This means that if the input data and the hashing algorithm used remain the same, the output will be the same. Hashing allows you to confirm that the data has not been altered. As an example, if you download files from the Kali Linux website (`https://www.kali.org/`), you will be able to check whether the hash of the file you have downloaded is the same hash as the original file. This helps you identify whether a malicious hacker might have altered the file from their website or if the file is corrupted.

What is the difference between asymmetric and symmetric encryption?

Symmetric encryption uses a single protected key that both the sender and receiver of the message know. An advantage of symmetric encryption is that it is faster and requires less computation power than asymmetric encryption.

In the following example, Alice wants to send an encrypted message to her friend, Bob. Alice encrypts the message with the shared private key, which converts (encrypts) the message from plaintext to ciphertext. When Bob receives the message, he uses the same shared private key to decrypt the message and read the plaintext message.

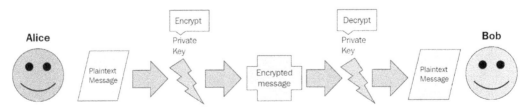

Figure 6.1: Symmetric encryption example

Asymmetric encryption uses a public and private key. The public key is shared by the owner, but their private key must only be known by them. Asymmetric encryption provides confidentiality through encryption and authenticity and non-repudiation through digital signatures.

In the following example, Alice encrypts her plaintext message with Bob's shared public key, and when Bob receives the message, he decrypts it with his private key (known only to him) to view the content of the message.

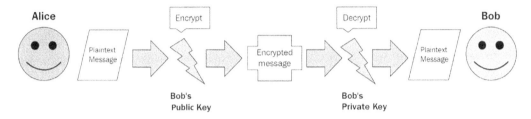

Figure 6.2: Asymmetric encryption example

What are some examples of symmetric encryption algorithms?

Here are some commonly used symmetric algorithms.

- **Advanced Encryption Standard (AES)**: AES is a widely adopted encryption standard that supports key sizes of 128, 192, and 256 bits. It is used in a variety of applications, including securing data at rest and data in transit, and in protocols like TLS and IPSec.

- **Twofish**: Twofish is a symmetric key block cipher that uses key sizes of up to 256 bits and is used in software encryption tools like TrueCrypt and GPG.

- **Serpent**: Serpent uses a block size of 128 bits and key sizes of 128, 192, or 256 bits. It is designed to be highly secure, with a slightly higher computational cost compared to AES. While not as widely adopted as AES, Serpent is used in some cryptographic applications that prioritize security over speed.

- **ChaCha20**: ChaCha20 is a stream cipher that was designed as an alternative to AES and is known for its speed and resistance to side-channel attacks. It is used in applications such as TLS (specifically in Google's QUIC protocol), OpenSSH, and encrypted messaging apps like Signal.

- **Blowfish**: Blowfish is a symmetric key block cipher that has a variable key length from 32 bits to 448 bits and is known for its speed and effectiveness. Blowfish is commonly used in cryptographic applications such as password management and file encryption..

What are some examples of asymmetric encryption algorithms?

Asymmetric algorithms that are used as of June, 2024, include the following::

- **RSA**: RSA is one of the most widely used public-key cryptographic algorithms. It relies on the computational difficulty of factoring large composite numbers. RSA is used for secure data transmission, digital signatures, and key exchange. It's commonly used in protocols like SSL/TLS for securing web traffic and in email encryption systems like PGP.

- **Elliptic Curve Cryptography (ECC)**: ECC is based on the algebraic structure of elliptic curves over finite fields. It offers the same level of security as RSA but with much smaller key sizes, resulting in faster computations and reduced storage requirements. ECC is widely used in mobile devices and constrained environments due to its efficiency. It's employed in protocols such as **ECDSA** (short for **Elliptic Curve Digital Signature Algorithm**), **ECDH** (short for **Elliptic Curve Diffie-Hellman**), and in securing communications in TLS.

- **Digital Signature Algorithm (DSA)**: DSA is a **Federal Information Processing Standard (FIPS)** for digital signatures. It is based on the mathematical concept of modular exponentiation and discrete logarithms. DSA is primarily used for digital signatures, ensuring data integrity and authenticity. It's often used in secure email systems, software distribution, and government applications.

- **ElGamal**: ElGamal encryption is based on the Diffie-Hellman key exchange. It provides semantic security and is often used for encryption and digital signatures. It's used in PGP/GPG for secure email encryption and in some blockchain technologies for secure transaction verification.

- **Paillier cryptosystem**: The Paillier cryptosystem is based on the composite residuosity class problem. It supports homomorphic encryption, allowing certain operations to be performed on ciphertexts, producing an encrypted result that matches the result of operations performed on the plaintexts. It's used in applications requiring homomorphic properties, such as secure voting systems, privacy-preserving data mining, and encrypted database queries.

What is the difference between RC4, RC5, and RC6?

RC4 is a stream cipher known for its simplicity and speed, encrypting data one byte at a time. It typically uses keys ranging from 40 to 256 bits and was historically used in SSL/TLS and WEP for its speed and simplicity, though it's now considered insecure due to vulnerabilities.

RC5 is a block cipher with a flexible design, featuring variable block sizes (32, 64, or 128 bits), key sizes (up to 2,040 bits), and rounds (up to 255). It uses addition, XOR, and bitwise rotation operations and is used in various applications for its flexibility but is less commonly used than AES.

RC6 is an advanced block cipher derived from RC5, featuring a fixed 128-bit block size, variable key sizes (128, 192, or 256 bits), and a fixed 20-round structure. It adds multiplication operations for increased diffusion and security.

What is AES and how does it work?

AES is a symmetric key block cipher standardized by NIST. It was designed to replace the older DES for enhanced security. AES encrypts data in 128-bit blocks using keys of 128, 192, or 256 bits, with the number of transformation rounds being 10, 12, or 14, respectively. The encryption process includes substitution with an S-box (SubBytes), shifting rows (ShiftRows), mixing columns (MixColumns), and adding a round key (AddRoundKey) derived from the original key via a key schedule, ensuring non-linearity and diffusion to thwart cryptographic attacks. AES is widely used in securing data transmission over the internet with the TLS protocol.

What is Triple Data Encryption Standard (3DES) and how does it work?

3DES is an encryption algorithm designed to enhance the security of the original DES by applying the DES cipher algorithm three times to each data block. It works by using three 56-bit keys, resulting in an effective key length of 168 bits. The process involves encrypting the data with the first key, decrypting it with the second key, and then encrypting it again with the third key. This triple-layer approach increases security compared to single DES. 3DES is sometimes used in securing financial transactions in the banking industry.

What is IDEA (International Data Encryption Algorithm) and how does it work?

IDEA is a symmetric key block cipher that encrypts data in 64-bit blocks using a 128-bit key. The encryption process involves eight rounds of complex transformations, each including operations such as XOR, addition, and multiplication modulo $2^{16} + 1$. These operations include mixing the bits of the plaintext and the key in a non-linear and non-reversible way.

IDEA is used in real-world applications like securing email communications with **Pretty Good Privacy (PGP)** encryption.

What is PGP and how does it work?

PGP is an encryption program that provides cryptographic privacy and authentication for data communication. It uses a combination of symmetric key encryption for speed and public key encryption for secure key exchange. When a message is encrypted with PGP, it is first compressed and then encrypted with a unique session key using a fast symmetric encryption algorithm. This session key is then encrypted with the recipient's public key and sent along with the encrypted message. The recipient uses their private key to decrypt the session key and then uses it to decrypt the message. PGP is widely used for securing emails.

What is the SHA-2 family of hashing algorithms?

The SHA-2 family of hashing algorithms includes SHA-224, SHA-256, SHA-384, and SHA-512, each producing hash values of different bit lengths: 224, 256, 384, and 512 bits. These cryptographic hash functions take an input and produce a fixed-size string of bytes, which is unique to each unique input, making them ideal for ensuring data integrity.

SHA-256 is used in the Bitcoin blockchain to secure and validate transactions. Similarly, SHA-2 algorithms are used in digital certificates and SSL/TLS protocols to protect data transmitted over the internet.

Can you name some common cryptography attacks?

Common cryptography attacks include known-plaintext attacks, chosen-plaintext attacks, cipher-text-only attacks, replay attacks (typically in the form of machine/man-in-the-middle attacks), and chosen-ciphertext attacks.

Common cryptography attacks include:

- **Brute force attacks**, where attackers try every possible key until the correct one is found.
- **Machine-in-the-middle attacks**, where attackers intercept and potentially alter communications between two parties without their knowledge. These attacks used to be called **man-in-the-middle (MitM)**, but that terminology is being phased out by the industry.
- **Replay attacks**, where attackers capture and resend a valid data transmission to trick the receiver into unauthorized actions.
- **Side-channel attacks**, which exploit physical implementation details such as timing, power consumption, or electromagnetic leaks to extract secrets.
- **Cryptanalysis attacks**, which aim to find weaknesses in the cryptographic algorithms themselves, such as differential cryptanalysis or linear cryptanalysis, to break the encryption without brute force.

What is PKI?

Public key infrastructure (**PKI**) is used to describe the policies, software, and other infrastructure needed to manage digital certificates and public key encryption.

PKI is best explained with a simple example. In the following figure, our user, **Bob**, requests a certificate from a **registration authority** (**RA**). The RA then helps to validate the identity of the user and sends a request to the **certificate authority** (**CA**) to create a user certificate and keys. Once the CA has created the certificate and keys, it sends it to the user. **Bob** then submits this certificate to our user, **Robyn**. **Robyn** contacts the CA to validate the certificate and Bob's identity. The CA then verifies that the certificate from Bob is valid, so Robyn now trusts Bob.

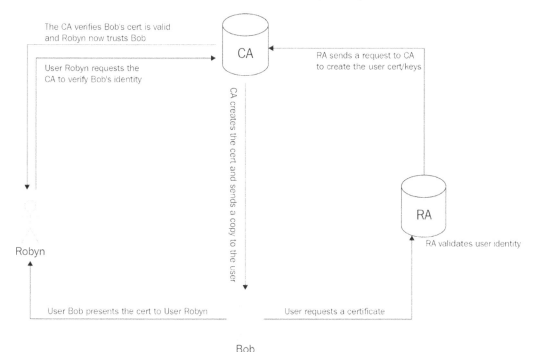

Figure 6.3: PKI example

Can you describe Quantum Cryptography?

Quantum cryptography uses light particles (photons) to transmit data between locations over fiber optic cables. The sender of the data transmits the light particles through a polarizer that characterizes the particles with one of the four possible bit polarizations:

- Horizontal (0-bit)
- Vertical (1-bit)

- 45 degrees left (0-bit)
- 45 degrees right (1-bit)

The particles then move to a receiver that uses two beam splitters to read the polarization of each particle. The receiver then notifies the sender of which beam splitter was used for the particles in the sequence that was sent; the sender compares that information with the sequence of polarizers that was used to send the key and discards any photons that were read with the wrong beam splitter. The remaining sequence of bits becomes the key. This method helps protect data because if anyone eavesdrops on the communication and reads or attempts to copy the data, the state of the photon will change and be detected by the endpoint. This is known as the Heisenberg uncertainty principle.

What is the difference between stream and block ciphers?

The primary difference between stream and block ciphers lies in how they process data.

Stream ciphers encrypt data one bit or byte at a time, making them suitable for scenarios where the data size is unknown or continuous, such as securing real-time communications (e.g., RC4 used in some SSL/TLS implementations).

Block ciphers encrypt data in fixed-size blocks, typically 64 or 128 bits, which makes them suitable for encrypting data with a defined size(e.g., files or database records) An example of using a block cipher is the use of AES to encrypt data at rest on a database.

A client uses RSA to encrypt their communication. What recommendations would you give the client to ensure their RSA implementation is secure?

Example answer

To ensure the security of their RSA implementation, I would recommend:

- Using a key length of at least 2,048 bits to guard against brute-force attacks
- Ensuring that proper padding schemes like **Optimal Asymmetric Encryption Padding (OAEP)** are used to prevent padding oracle attacks
- Regularly rotating keys to limit exposure if a key is compromised

- Implementing secure storage and handling procedures for private keys

- Using strong random number generators for key generation to avoid predictable keys

During a security review, you discover that an application is using MD5 to hash passwords. What steps would you take?

Example answer

I would immediately recommend migrating to a more secure hashing algorithm, such as bcrypt, scrypt, or Argon2, which are specifically designed for password hashing with built-in salting and stretching mechanisms. I would also recommend implementing a process to rehash existing passwords using the new algorithm as users log in and educating the team on the importance of using secure hashing methods.

You are tasked with securing a communications protocol. What cryptographic methods would you use and why?

This question is a bit generic in nature, so here is an example response.

I would use a combination of symmetric encryption (like AES) for data confidentiality, asymmetric encryption (like RSA) for key exchange, and hashing algorithms (like SHA-256) for integrity checks. Symmetric encryption (AES) is efficient for encrypting data streams, asymmetric encryption (RSA) ensures secure key exchange over insecure channels, and hashing (SHA-256) provides a means to verify data integrity and authenticity.

How would you secure data at rest located in a database?

Example answer

To secure data at rest in a database, I would do the following:

- Use strong encryption algorithms like AES-256 to encrypt sensitive data.

- Implement column-level encryption for highly sensitive fields.

- Ensure that encryption keys are securely managed using a **key management service** (KMS).

- Implement access controls and auditing to monitor who accesses the encrypted data.

- Ensure that backups are also encrypted and securely stored.

How would you perform a security audit on a cryptographic implementation?

Other teams in the organization might be involved in some of the items (e.g., penetration testing) listed below.

Example answer

To perform a security audit on a cryptographic implementation, I would do the following:

- Review the algorithm and protocol specifications to ensure they are up to date with current standards.
- Examine the source code for proper implementation, looking for common vulnerabilities like improper key management, lack of salting, or incorrect usage of cryptographic primitives.
- Perform tests to check for side-channel vulnerabilities and timing attacks.
- Validate the randomness of key and nonce generation.
- Conduct penetration testing to see if any known attacks can be applied successfully.

Another member of the team asks you to explain the concept of forward secrecy. How would you describe it?

Forward secrecy, also known as **perfect forward secrecy** (**PFS**), ensures that session keys used for encrypted communications are not compromised even if the server's private key is compromised in the future. This is typically achieved by generating unique session keys for each communication session using ephemeral key exchange mechanisms (e.g., **Diffie-Hellman Ephemeral** (**DHE**) or **Elliptic Curve Diffie-Hellman Ephemeral** (**ECDHE**)). This way, even if long-term keys are compromised, past communications remain secure.

How would you approach decrypting a message if you do not have the decryption key?

Without the decryption key, I would analyze the encryption method used. If it is a weak or flawed implementation, cryptanalysis techniques like known-plaintext attacks, chosen-plaintext attacks, or brute force might be applicable.

However, if the encryption is strong and properly implemented, decrypting the message without the key would be infeasible, and I would seek alternative methods, such as obtaining the key through other means (e.g., a social engineering attack or exploiting system vulnerabilities).

A client uses SSL/TLS for securing web traffic. What best practices would you recommend?

As of June, 2024, these would be recommendations I would provide the client.

Example answer

For securing web traffic with SSL/TLS, I would recommend:

- Using the latest version of TLS to leverage its security improvements and performance benefits
- Disabling older, vulnerable protocols
- Using strong cipher suites that support forward secrecy
- Regularly updating certificates and using **Extended Validation** (**EV**) certificates where appropriate
- Enabling **HSTS** (**HTTP Strict Transport Security**) to prevent downgrade attacks

How would you handle key management for an encryption system?

As of June, 2024, of writing, these are my recommendations.

Example answer

For key management, I would:

- Use a secure KMS for generating, storing, and rotating keys.
- Implement strict access controls to ensure that only authorized personnel can access the keys.
- Use **hardware security modules** (**HSMs**) to protect keys in a secure hardware environment.
- Regularly audit key access and usage.
- Ensure that key backup procedures are in place and that backups are securely stored and encrypted.

Summary

In this chapter, you learned about the cryptography/cryptanalyst career path and the average salary for this career in the US. You also learned how working as a cryptographer can be a stepping stone into more advanced careers and learned some of the common interview questions that might be asked of you.

Just a reminder: the end of this book contains a list of salary checking resources, so you can find salary information for your location in the world.

In the next chapter, you will learn about a career as a GRC analyst, including common knowledge-based interview questions you might be asked.

Join us on Discord!

Read this book alongside other users. Ask questions, provide solutions to other readers, and much more.

Scan the QR code or visit the link to join the community.

`https://packt.link/SecNet`

7

GRC/Privacy Analyst

This chapter focuses on the **governance, risk, and compliance (GRC)** analyst and privacy analyst roles. These roles are commonly overlooked as cybersecurity roles, given their focus on policies and procedures rather than hands-on keyboard technical skills. Still, they are a great path into the world of cybersecurity and provide exposure to many different areas of the field.

The following topics will be covered in this chapter:

- What is a GRC/privacy analyst?
- How much can you make in this career?
- What other careers can you pursue?
- What certifications should be considered?
- Common interview questions for a GRC/privacy analyst

What is a GRC/privacy analyst?

A GRC analyst is responsible for assessing and documenting an organization's policies and regulations associated with the compliance and risk posture of information assets. An individual in this role aims to ensure alignment between technology decisions and business outcomes while improving operational efficiencies. GRC analysts may function as part of an organization's internal security team or in more of a consultative capacity in which they would guide a company's internal security team as a representative of a third-party organization.

Specific responsibilities may include risk identification and analysis, policy development, auditing, reporting, and regulatory compliance. A privacy analyst's role is similar to that of a GRC analyst from an activity and responsibility perspective.

The difference, however, is that a privacy analyst, as the name implies, is specifically focused on ensuring that a company's business operations, policies, and procedures meet privacy requirements and regulations pertaining to protecting critical information. GRC and privacy analysts collaborate with various departments, participate in incident response, and are responsible for continuous compliance monitoring. Some of the applicable laws for privacy analysts include the **General Data Protection Regulation (GDPR)**, which is a **European Union (EU)** regulation on information privacy in the EU and the European Economic Area; the **California Consumer Privacy Act (CCPA)** is a state statute intended to enhance privacy rights and consumer protection for residents of the state of California in the United States, as well as other states developing their own varying privacy laws.

How much can you make in this career?

As of June, 2024, the average annual salary for both GRC and privacy analysts falls within the range of $81,000 to $126,000. Salary influencers for this type of role are based on experience level, specialization, industry, and location in the US.

What other careers can you pursue?

GRC/privacy analysts require a broad perspective on corporate functions and processes. Although the role is not directly responsible for the management and deployment of technology, individuals in this role must possess a solid understanding of risk management and the technology used as mitigating controls, thereby making the possibilities endless.

In addition, GRC/privacy analysts can potentially be exposed to companies that span a broad range of industries, which, in turn, may lead to working with various laws and regulatory requirements, providing a path to focusing on a career as a Security Auditor (see *Chapter 8*, *Security Auditor*, for more details).

From a soft skills perspective, constant effective communication (both written and verbal) is required for this type of role due to interviewing and documentation requirements. This experience can help build confidence and the skills to lead and manage teams.

What certifications should be considered?

To get started in the role of a GRC/privacy analyst, you can consider the following courses:

- **ISACA's Certified in Risk and Information Systems Control (CRISC)** is meant for professionals who deal with information systems risks. IT, audit, risk, and cybersecurity professionals usually pursue the certification program during their mid-senior stage. For more information, visit https://www.isaca.org/credentialing/crisc.

- **ISACA's Certified Information Security Auditor (CISA)** is suited for professional and experienced auditing, monitoring, controlling, and assessing organizations' business and IT systems. For more information, visit `https://www.isaca.org/credentialing/cisa`.

- **GRC Professional (GRCP)** is issued by the **Open Compliance and Ethics Group (OCEG)** and demonstrates an individual's understanding of GRC. It can be pursued by individuals at various stages of their careers, whether they are starting in an auditing role or are already GRC practitioners. For more information, visit `https://www.oceg.org/certifications/grc-professional-certification/`.

Next, take a look at a few interview questions that will help you with your interview.

Common interview questions for a GRC/privacy analyst

As a GRC/privacy analyst, it is essential to understand common frameworks, standards, and regulations relevant to the industry vertical of the company being pursued as an employer. For example, knowledge of the **Health Insurance Portability and Accountability Act (HIPAA)** and HITRUST (`https://hitrustalliance.net/`) would be necessary for an analyst pursuing a career in the healthcare industry.

At the same time, PCI-DSS, **Sarbanes-Oxley (SOX)** would be more suited for an analyst headed down the finance path. For those advising on data and privacy-related roles, GDPR, CCPA, and other data privacy and sovereignty laws would be of concern.

The following is a list of interview questions that could prove helpful in preparing for a GRC/privacy analyst interview:

What is GRC and why is it essential to an organization?

Seems like a silly question, right? Well, has anyone ever asked you your age, and you had to stop and think about it for a moment? You don't want that to happen to you in an interview. A straightforward question, and sure, you know exactly what it is in your mind, but have you practiced articulating it verbally?

Let's start with the easy part. GRC is a strategy used by organizations to manage **governance** (ensuring IT processes align with business goals), **risk** (events or situations that could be damaging to the organization), and **compliance** (adherence to legal and regulatory requirements). GRC is essential to a company's success and sustainability as it directly impacts an organization's ability to achieve its goals and align with stakeholders' desired outcomes.

Benefits include the following:

- It improves activities associated with managing, identifying, and evaluating risk.

- It aids in strategic planning related to policy and corporate management activities.

- It focuses on adherence to legal and regulatory compliance requirements.

Now, let's dig a bit more into risk management.

Why is risk management an important component of an organization's governance plan?

Risk management aims to inform the development of strategic security initiatives aimed at identifying and evaluating organizational risk and then reducing the risk to an acceptable level. Sources of corporate risk are not limited to technology, however. When assessing organizational risk, it is essential to consider all aspects of risk (logical, physical, internal, external, and so on) so that the organization is not vulnerable due to blind spots. For example, when evaluating a manufacturer's organizational risk, analysts should consider intellectual property, data privacy, and critical infrastructure.

After identifying and evaluating the risk posed to an organization, a decision will need to be made regarding the appropriate action to take in response; this informs the risk mitigation strategy. The four most commonly exercised risk responses are as follows:

- **Risk acceptance**: When a risk is accepted, security leaders and business decision-makers have agreed to own the consequences and loss associated with a realized risk. For example, a company may elect to accept an identified risk because the cost associated with the loss does not outweigh the countermeasure cost.

- **Risk transfer**: Companies transfer risk by assigning ownership of the cost of a loss to a separate entity. Examples of this include purchasing cybersecurity insurance and outsourcing.

- **Risk avoidance**: Avoiding risk involves replacing it with a less risky alternative or removing it altogether. An example is building a data center in a location other than Florida to avoid hurricanes. Another example is avoiding **File Transfer Protocol** (**FTP**) attacks by turning off FTP across the organization.

- **Risk reduction/mitigation**: The goal of risk reduction is to implement controls and countermeasures to protect the organization from threats and eradicate vulnerabilities. The approach should be cost effective based on the cost/benefit analysis performed during risk identification and evaluation.

How can GRC and cybersecurity teams work together?

The risk management process should lead to risk-reducing outcomes, which aligns with the need to communicate security program effectiveness in risk-reducing terms. Security teams find it increasingly difficult to identify, monitor, and measure organizational risk due to complex architecture, a lack of standardized information system policies, and the uncertainties of unknown security threats and unpredictable user behavior.

Maturity-based and cyber risk-based (or threat-focused) are the two most common approaches to cybersecurity. While both processes offer the ability to monitor and measure program effectiveness, they produce very distinct outcomes.

How can maturity-based approaches help with a GRC program?

Maturity-based programs focus on building defined capabilities to achieve a desired level of maturity. Although the technique does not explicitly target risk reduction as an outcome, it is still considered the industry norm for cyber risk management. While organizations seeking to establish an operation have seen success with this approach, as the model matures, the constant requirement to monitor and improve all program capabilities often results in inefficient spending and unmanageable tool sprawl.

How can a cyber risk-based approach help the GRC program?

This approach is more cyber technical than the risk-based approach and requires a significant cyber maturity of the teams of analysts. A cyber risk-based focus positions companies to identify the most critical vulnerabilities and threats and implement countermeasures to reduce the most crucial organizational cyber risks. This path lets security teams prioritize program investments (driving cost-effective decisions) and articulate risk-reducing outcomes. Many industry experts argue that when compared to the maturity-based approach, it is the cyber risk-based operating model that is better suited to help security teams keep pace with the evolving threat landscape due to its flexibility and specificity. However, the inherent challenge for security teams when leveraging this approach is the lack of well-defined metrics due to the wide range of actions companies may prioritize.

What is CIA and why is it important?

In the context of cybersecurity, CIA refers to the confidentially, integrity, and availability of data and systems. The CIA triad serves as the foundation or base-level principle for the development of security policies and systems. Confidentiality focuses on ensuring that data can only be viewed by authorized entities.

The goal of integrity is to guarantee that data remains tamper-free and trustworthy. Integrity is lost if data is not authentic, accurate, and reliable. If data is inaccessible, the importance of confidentiality and integrity diminishes somewhat. Availability efforts aim to ensure that data, applications, systems, networks, and so on are available and function properly.

How can GRC analysts help with data governance and data? What are some of the common types of frameworks used?

Generally, frameworks enable business leaders and professionals to communicate in a common language, leveraging standards. Similarly, cybersecurity frameworks aid security leaders in securing digital assets by providing a reliable and logical process for reducing cyber risk. It is important to note that there are three types of security frameworks:

- **Control frameworks** offer guidance on assessing the state of an organization's technical controls, establishing a baseline of technical controls deemed appropriate for the business, and prioritizing implementation. The Center for Internet Security's Critical Security Controls is a widely used control framework. The following is a list of the CIS controls in v8 and how many safeguards in each apply to each of the control implementation groups:

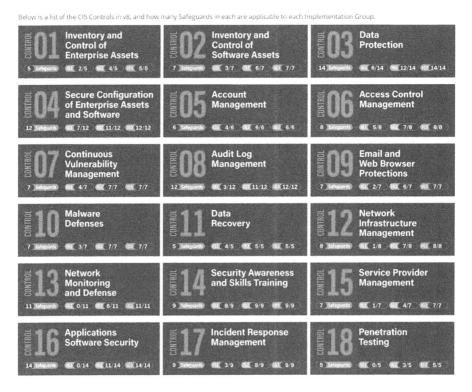

Figure 7.1: CIS controls in v8 (source: https://www.cisecurity.org/controls)

- **Program frameworks** provide a process for assessing, building, and measuring an organization's security program. An example of a program framework is the **National Institute of Standards and Technology (NIST)** framework for improving critical infrastructure cybersecurity (NIST Cybersecurity Framework). The following diagram shows version 2.0 of the NIST CSF:

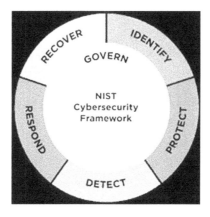

Figure 7.2: NIST Cybersecurity Framework 2.0

- **Risk frameworks** offer a structured approach to managing cyber risk by identifying, measuring, and quantifying risk and prioritizing mitigation activities. The NIST **Risk Management Framework (RMF)** is a common framework that governments and contractors use to manage and demonstrate cyber risk mitigation activities and security posture improvements. The following diagram is a snapshot of the NIST RMF and the definitions of each of the stages:

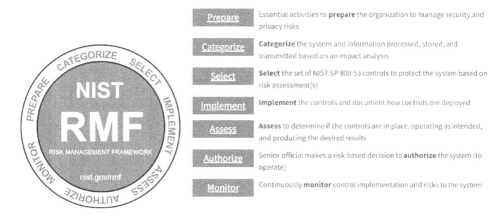

Figure 7.3: NIST Risk Management Framework
(source: https://csrc.nist.gov/Projects/risk-management)

How does preventive control differ from detective control?

As the names imply, the purpose of preventive control is to actively prevent a threat, while detective control will only detect a threat. You might ask yourself why an organization would want to use a control only capable of detecting but not capable of taking any action, and that's a great question. It boils down to taking a layered or defense-in-depth approach to security and using the right tool for the job.

A CCTV system is an example of a detective control, as it is usually used after the fact to review footage of an event. Roving guards and gates act as a preventative control for buildings.

For another example, consider modern-day **endpoint protection** (**EPP**), which typically offers a combination of a **next-generation antivirus** (**NGAV**) and **endpoint detection and response** (**EDR**). Together, they are compelling because the NGAV will prevent malicious activity, and the EDR records all of the process-level activity so that in the event a threat can bypass the NGAV, the activity can still be detected, and detailed forensics are available to assist in determining the appropriate response action. The following diagram aligns the different control types to their place in the life cycle of a cyber attack:

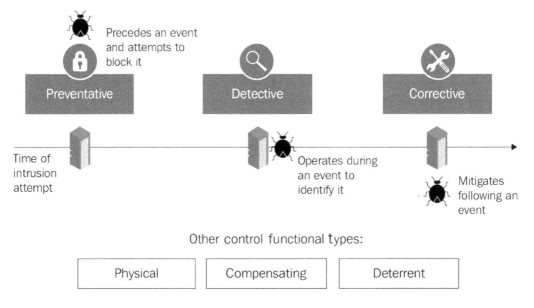

Figure 7.4: Control types

Describe how GRC risk management is used

GRC risk management aims to identify, classify, and quantify organizational risk. This phase requires appropriate guidance from organizational management on the guidelines for defining an organization's critical mission processes, its critical assets, and impacts on those assets or processes if they were to be negatively impacted.

As analysts review the various threats to organizational assets or business processes, they must determine the potential impact on the organization depending on the various situations. It is important to classify the risks to the organization in ways that management can make meaningful. There are qualitative ways (high, medium, low) and quantitative ways (FAIR™ quantitative risk analysis model: `https://www.fairinstitute.org/fair-risk-management`) to communicate this risk to management depending on the ability to accurately describe the impact of the risk on the business.

As part of the process, the appropriate risk responses (avoid, mitigate, accept, transfer) are determined, and, in the case of mitigation, preventive and detective mitigation control methods are performed. Examples of this would include exiting the line of business to avoid the risk, installing controls to help limit or mitigate this risk, choosing to do nothing as the risk to the business is not worth the cost of the controls, or getting an insurance policy to transfer the financial cost of the risk to the insurance company.

How is risk calculated?

Risk is defined as the potential for damage, the loss of data, or the destruction of assets caused by the combination of a threat and a vulnerability to an organizational asset. It can be further refined by the impact on the organization as a multiple of the incident's likelihood.

A vulnerability represents a weakness in an organization's environment. This weakness can exist in software, hardware, and even in processes. A threat, on the other hand, is anything that can exploit a vulnerability.

The terms threat, vulnerability, and risk are often confused, so be sure you're clear on each term as a standalone entity and concerning one another.

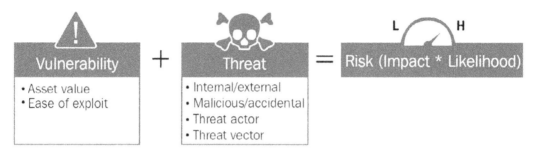

Figure 7.5: Risk formula

How should a risk be classified?

This question can be a trap. The answer is straightforward, but at the right moment, it could lead a candidate down a long and dark path of overexplaining. The simple answer is that while risk should ultimately be classified by its qualitative or quantitative rating, the specifics of the classifications should align with the company's risk policy.

Define and provide examples of risk response within the context of risk management.

When managing an organization's cybersecurity risk, the overall outcome is to determine a cost-effective solution for reducing or mitigating the risk to an acceptable level according to the organization's priorities and tolerance levels. The four most common responses are as follows:

- **Accept**: This typically means the cost of mitigation outweighs the cost of risk realization.
- **Mitigate/reduce**: Implement controls to lessen the risk to a level within the organization's tolerance level.
- **Transfer**: Make it someone else's problem (in other words, cybersecurity insurance).
- **Avoid**: The line of business or process should be stopped as it presents a significant risk to the company.

What makes a successful GRC/privacy analyst?

In addition to understanding the necessary framework, methodologies, regulatory requirements, and more, successful GRC/privacy analysts can build relationships and communicate with various teams with varying levels of technical and business acumen. The analyst must be able to serve as a bridge and interpreter between cross-functional teams and different organizational units.

Depending on the organization and the role, an analyst may also need to analyze data, produce reports, parse through new regulations, put them in the context of the org, and make recommendations to management.

Provide an example of a risk management framework

A popular and widely adopted risk management framework is NIST RMF (`https://csrc.nist.gov/projects/risk-management/about-rmf`). According to NIST, this framework provides a method for integrating security, privacy, and cyber supply chain risk management activities into the system development life cycle. The goal of the framework is to aid in selecting and integrating security controls in a risk-based fashion. The steps of the framework are as follows: Prepare, Categorize, Select, Implement, Assess, Authorize, and Monitor.

What is GDPR, and when does it apply?

GDPR is viewed as the world's most challenging privacy and security law. The regulation's effective date was May 25, 2018, and although it was drafted and passed by the **EU**, its impact is global.

The regulation is intended to protect the privacy and security of data and targets any organization in the EU that handles data related to people.

Other applicable privacy-related laws and regulations include the US Privacy Act, CCPA, and US state data privacy laws. Some notable international privacy laws include the **Canada Privacy Protection Act (CPPA)**, the New Zealand Privacy Act, and privacy and data protection laws from Brazil, Thailand, Singapore, China, Switzerland, and more.

What other privacy laws are you familiar with and what industries are they applicable to?

These laws include the following:

- **HIPAA**: This standard governs **patient health information (PHI)** protection and applies to the healthcare industry.
- **Family Educational Rights and Privacy Act (FERPA)**: This federal law applies to all schools receiving funds through a US Department of Education program. The law protects the privacy of student education records.
- **GDPR**: This protects the privacy and security of people's data in the EU. It does not apply to a specific industry. The regulation has a broad reach and applies to any organization in any industry that handles data related to people in the EU.

- **Data Protection Framework (DPF):** On July 17, 2023, the European Commission issued an adequacy decision on the EU-US **DPF**. This new voluntary framework, which replaces the Privacy Shield program, provides a mechanism for companies to transfer personal data from the EU to the US in a privacy-protective way consistent with EU law (`https://www.ftc.gov/business-guidance/privacy-security/data-privacy-framework`).

Also of note is that three states have introduced consumer data privacy laws:

- **CPA: Colorado's Privacy Act** (which was created within its Consumer Protection Act)
- Virginia's Consumer Data Protection Act

At a high level, the laws are in place to provide consumers with the right to access, delete, update, and/or obtain a copy of their data. There are some slight differences within the laws, so having a solid understanding of any laws that apply to your potential employer would be worthwhile.

Additional details on these and other essential consumer privacy laws can be found on the **National Conference of State Legislatures (NCSL)** website: `https://www.ncsl.org/`.

What is data sovereignty?

Data sovereignty is the idea that data is regulated by the laws of the country or region in which it is processed or used. Another term often combined with sovereignty is data localization, which is storing data within the physical boundaries of the country or region where it originated. These practices are used in combination because data is then processed and stored within the same regional or country boundaries and its applicable laws, simplifying the management process.

What is DLM?

One of the tasks that more and more GRC analysts are being asked to participate in is the governance and life cycle of data management. DLM is an approach to managing data from the time of entry to its destruction and throughout its life cycle.

Figure 7.6: DLM
(source: https://www.clicdata.com/blog/complete-guide-data-lifecycle-management/)

Within the data life cycle management framework, governance ensures the data is secure, accurate, private, useable, and available. This applies to all the people who must use and process the data and the technology that enables it throughout its life.

While every organization has unique goals, needs, and structure, here are the four most common data governance roles:

- Data admin
- Data steward
- Data custodian
- Data user

These are usually labels assigned to different users within the organization, who are responsible for managing the data and its usage. Additional resources for this include the **Data Management Body of Knowledge (DMBOK)** by DAMA International and Atlan (`https://atlan.com/data-governance-pillars/`) or a higher-level dive into the topic.

What is the difference between security and privacy?

Although the two are very closely related, they are not the same. Privacy refers to an individual's right to control the use of their personal information, whereas security refers to how data is protected: use versus protection. There is a clear overlap between the two, but they are still two separate and essential considerations. The following scenario illustrates a few examples of the differences.

Scenario: You are admitted to the hospital, and they need to access your medical records from your primary care and pharmacy. How does the hospital get access to your information from your medical providers? Let's see:

- The HIPAA Privacy Rule aims to ensure that PHI remains protected and secure while enabling the flow of health information. Signing your HIPAA privacy release at the hospital permits them to request medical information from your doctor and pharmacy.

  Outcome: *Your information is safe and your privacy is maintained.*

- The HIPAA Security Rule requires covered entities and business associates to implement technical, physical, and administrative safeguards. These safeguards ensure the security of your data as it traverses the internet and between organizations.

  Outcome: *Your information is safe in transit and protected from outside actors looking at it.*

Scenario: You walk into a car dealership and purchase a car. The dealer collects your personal information and stores your payment details. What's the worst that could happen? Let's take a look:

- The dealer uses your personal information to open a loan with the bank to cover the cost of your vehicle. They charge the card you left on file $2,500 to cover the down payment. Your data is stored on a secure server and only accessible to individuals who require access to perform their daily jobs.

  Outcome: *Your information is safe, and your privacy is maintained.*

- You signed a bunch of paperwork when you were there, and one of the documents was the dealer's privacy disclosure. The dealer sells your information to an agency marketing services to significant auto insurance companies.

  Outcome: *Your information is still secure, but your privacy is now compromised.*

- The dealership is breached, and attackers steal the account details of everyone who has ever considered purchasing a car from this location since its doors opened 15 years ago.

  Outcome: *The security and the privacy of your data are compromised.*

Explain how encryption can be leveraged to ensure data privacy

Data privacy refers to an individual's right to control the use of their personal information. Encryption makes this possible by scrambling plain text into an unreadable format with an encryption key. This hides the data from potential eavesdroppers because only the person or entity with the decryption key can convert the unreadable data into plain text.

How does technical control differ from administrative control?

Using the preceding examples, encryption would be an example of a technical control used to protect data in transit from being read by an outside observer. In contrast, an administrative control is a policy or adherence to a regulation denoting who might have permission to view protected health information.

The preceding questions are intended to offer insights into the types of questions you might be asked in an interview for a GRC/privacy analyst role. You should better understand how to narrow your focus in preparation for interviews of this type.

 It is not enough to simply understand the history and critical points within laws and regulations. It is also essential to comprehend the problem being solved and the potential business impact and implications. This will vary from organization to organization and from industry to industry.

Lastly, a great way to differentiate yourself in an interview is to be aware and well informed of upcoming changes to laws and regulations, especially those that directly impact the organization that has called you in for an interview. For example, the CCPA expands California's consumer data privacy laws and it took effect on January 1, 2023.

Summary

In this chapter, you learned what a GRC/privacy analyst is and the average salaries in the US for roles of this type, along with certification considerations, career path options, and common questions you might be asked during an interview. While understanding the questions and answers as stated is essential, explore beyond the specifics in the questions themselves to ensure you know the surrounding concepts.

In the next chapter, we will learn about a career as a Security Auditor.

Join us on Discord!

Read this book alongside other users. Ask questions, provide solutions to other readers, and much more.

Scan the QR code or visit the link to join the community.

https://packt.link/SecNet

8

Security Auditor

In this chapter, you will learn what a Security Auditor is and the average salary range for this career in the US. You will also learn about the career progression options and common interview questions for the role.

The following topics will be covered in this chapter:

- What is a Security Auditor?
- How much can you make in this career?
- What other careers can you pursue?
- Common interview questions for a Security Auditor interview

What is a Security Auditor?

A Security Auditor is an individual who helps to provide an independent systematic review of an organization's information security system. Sometimes, they work as individuals. Other times, they can perform as part of a team or department providing audit services inside an organization. Security auditors can also be external consultants who provide an independent systematic review of their client's information security system or scoped parts per their contract.

Security auditors conduct their audits based on organizational policies and any applicable government compliance and regulations. They work with **information technology (IT)** personnel, security, managers, executives, and other business stakeholders to validate the business's industry best practices versus any applicable policy regulation or best practice. Auditors achieve this by using questionnaires and interviews, monitoring the process flows/work actions of the companies they are auditing, examining samples of past activities or logs, or validating that controls and procedures work versus how they are expected to work. This sometimes requires on-site assessment and validation of technical or physical security controls.

How much can you make in this career?

The salary range of a Security Auditor ranges from $60,000 to $130,000 USD. It can be higher or lower depending on the candidate's location in the US, years of experience, and demand for the other areas of specialization that they might have, for example, cloud security or application security experience.

What other careers can you pursue?

Having a career as a Security Auditor prepares you for various other occupations. The combination of experiences you gain as a Security Auditor allows you to achieve the expertise needed to provide consulting engagements. For example, as a Security Auditor, you regularly look at best practices and make recommendations for testing and control programs. Security auditors often go on to help lead **governance, risk, and compliance** (GRC) programs for organizations, using their auditor experience to highlight and remediate deficiencies in making laws, regulations, and internal requirements.

Other subspecialties include secure software assessors or security control assessors, who might be more prominent in the government or government contractor space.

Common interview questions for a Security Auditor interview

The following is a list of interview questions that could prove useful in preparing for a Security Auditor interview.

How do you stay updated with the latest security threats and compliance regulations?

Keeping up with the latest security threats and compliance regulations can be as simple as setting up Google alerts for specific trigger words related to the systems or areas of concern in your specialization. Another recommendation could be using RSS feed aggregators to combine the RSS feeds from various industries, news outlets, blogs, or even social media keywords that might be trending.

From there, you need to dig into the details to understand the impacts on the organization, why or how it can happen to an organization, and any applicable controls or procedures to follow as potential ways to prevent or minimize the impacts on the organization.

What frameworks are you familiar with or have you performed assessments against?

In addition to the internal policies and procedures of the hiring company, auditors will need to be familiar with federal regulations such as the **Health Insurance Portability and Accountability Act (HIPAA)** (`https://www.cdc.gov/phlp/php/resources/health-insurance-portability-and-accountability-act-of-1996-hipaa.html`) and **Sarbanes-Oxley Act (SOX)** (`https://www.law.cornell.edu/wex/sarbanes-oxley_act`), or standards set by the **International Organization for Standardization (ISO)** (`https://www.iso.org/isoiec-27001-information-security.html`) or the **National Institute for Standards in Technology (NIST)** (`https://www.nist.gov/`).

What are the standard certifications that a Security Auditor might have?

Some of the standard certifications that a Security Auditor might have are as follows:

- **Certified Information Systems Auditor (CISA)** from ISACA (`https://www.isaca.org/credentialing/cisa`).
- **Certified in Risk and Information Systems Control (CRISC)** from ISACA (`https://www.isaca.org/credentialing/crisc`) is helpful for those focusing on risk and system controls.
- **CompTIA Security+** from CompTIA (`https://www.comptia.org/certifications/security`). CySA+ is another helpful certification from this provider.
- **IT Infrastructure Library (ITIL)** certification (`https://niccs.cisa.gov/training/search/standard-technology-incorporated/it-infrastructure-library-itil-foundation`).
- Certifications for specific **cloud service providers (CSPs)** or systems providers.

While there are many other certification providers and certificates, I recommend you consider how a specific certificate is focused on your career and its trajectory.

How would you describe the difference between controls and control frameworks?

In the context of cybersecurity, "controls" and "control frameworks" are two related but distinct concepts.

Understanding their differences is crucial for effectively communicating with business stake-holders in cybersecurity:

- **Controls**: Controls are specific measures or mechanisms implemented to manage risks to an organization's information and systems. They are the actual tools, policies, or practices used to protect against threats and vulnerabilities. Controls can be of various types, such as technical, administrative, or physical. Controls are the individual building blocks that help in safeguarding an organization's assets.

  For example, imagine your organization's network as a house. In this analogy, a control would be like a lock on the door – it's a specific measure you put in place to prevent un-authorized access.

- **Control frameworks**: Control frameworks, on the other hand, are structured sets of guide-lines or best practices that help organizations in designing, implementing, maintaining, and evaluating their control systems. These frameworks provide a comprehensive ap-proach to managing cybersecurity risks by outlining how controls should be organized and how their effectiveness can be measured. They often come with industry-standard practices and are designed to help organizations achieve specific cybersecurity objectives or comply with regulatory requirements.

  Going back to the house analogy, if controls are locks, then a control framework would be the security plan for the house. It doesn't just tell you to have a lock but it also guides you on where the locks should be, what type of locks to use, how to maintain them, and how to ensure they are effective in keeping your house secure.

A well-known example of a control framework is the NIST Cybersecurity Framework `https://www.nist.gov/cyberframework`, which provides guidelines on managing and reducing cyber-security risks.

What are the differences between general, system, and application controls?

As an organization develops its internal policies and guidelines, they form the basis of general controls to which all the people and processes must adhere. For example, all procurement of system hardware needs to be done via company-approved vendors.

Controls then become more granular at the system level; company websites must be developed to be compatible with mobile and modern web browsers.

To achieve even more granularity, each system might have specific controls to meet internal policies and procedures. For example, company secrets must not be hardcoded into any published or shared code.

How would you approach an audit for an organization?

When coming into a new organization for an audit engagement, the lead must schedule a meeting with all the stakeholders to define the audit objectives and understand the organization's context. Work with the stakeholders to develop a preliminary schedule for the audit and all the potential individuals involved. Create a specific budget (even if you are not responsible for the financials, budget in time using hours) and define the scope of the engagement. Then, based on the budget or time and scope, you will list the audit team members, specify tasks for each individual, and determine the deadlines.

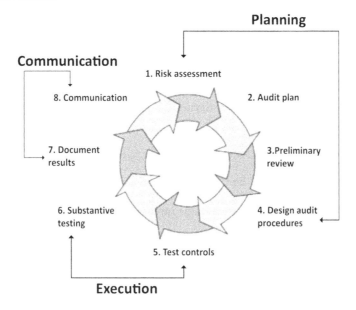

Figure 8.1: Audit Phases

With the plan in place, you will start the audit. An audit generally has the following eight phases:

1. Risk assessment
2. Audit plan
3. Preliminary review
4. Design audit procedures
5. Test controls
6. Substantive testing
7. Document results
8. Communication

More information on assessments and auditing can be found at `https://www.nist.gov/cyberframework/assessment-auditing-resources`. When working with or for the government, it is essential to follow the NIST RMF (`https://csrc.nist.gov/projects/risk-management`) and the federal controls (`https://csrc.nist.gov/Projects/risk-management/sp800-53-controls/release-search#!/800-53?version=4.0`).

Can you describe the phases of an audit?

There are three main phases of an audit, which are the planning phase, the execution phase, and the reporting phase:

- The **planning phase** contains the following steps:

    1. Risk assessment and determining the physical location that will be audited
    2. Determining the objective of the audit
    3. Determining the scope of the audit
    4. Pre-audit planning
    5. Determining the audit procedures that will be followed

- The **execution phase** contains the following steps:

    1. Gathering relevant data and documents to conduct the audit
    2. Evaluating existing controls to determine their effectiveness and efficiency
    3. Validating and documenting your observations during the audit and providing evidence

- The **reporting phase** contains the following steps:

    1. Creating a draft report and discussing it with management
    2. Issuing a final audit report that contains the findings of the audit, evidence, recommendations, comments from management, and the expected date of closure of the audit findings
    3. Conducting a follow-up to determine whether the audit findings are now closed and issuing a follow-up report

How does a Security Auditor validate that a control is functioning properly?

A Security Auditor plays a critical role in ensuring that cybersecurity controls are functioning effectively within an organization. The validation process typically involves several key steps:

1. **Review of Documentation:** The auditor begins by reviewing the documentation related to the specific control. This includes policies, procedures, and configuration settings. For example, if the control is a firewall, the auditor would examine the firewall configuration and the policies governing its use. Documentation review helps auditors understand how the control is supposed to function and the standards it should meet.

2. **Interviews and Observation:** Auditors often conduct interviews with staff responsible for implementing and maintaining the controls. These interviews provide insights into how the controls operate in practice and any challenges encountered. Additionally, observing the control in operation can help auditors to assess its practical effectiveness.

3. **Testing and Analysis:** This is a critical step where auditors perform various tests to validate the control's effectiveness. These tests can be:

    a. **Automated Testing:** Using tools to simulate attacks or to check for vulnerabilities

    b. **Manual Testing:** Manually checking systems and processes to ensure that controls are functioning as intended

    c. **Sampling:** Selecting a representative sample of data or transactions to check for compliance with control requirements

4. **Comparing Against Standards and Benchmarks:** Auditors compare the findings from their tests against established standards, benchmarks, or best practices. This comparison helps to determine whether the control meets the required cybersecurity standards.

5. **Reviewing Past Incidents:** Examining past security incidents or breaches can provide valuable information on the effectiveness of controls. If a control has failed in the past, the auditor investigates whether appropriate measures have been taken to address the failure.

6. **Reporting and Recommendations:** After the evaluation, auditors prepare a report detailing their findings, including any deficiencies or areas for improvement. They also provide recommendations for enhancing the control's effectiveness.

7. **Follow-up:** In some cases, auditors might perform follow-up audits to ensure that their recommendations have been implemented and that the control is now functioning correctly.

For instance, to validate a control like regular software updates, an auditor might review the update policy, interview IT staff responsible for updates, manually check a sample of systems to ensure they are up to date, and compare findings with best practice standards, like those from NIST.

Throughout this process, auditors use a combination of technical expertise, analytical skills, and an understanding of cybersecurity best practices to ensure that controls are not only in place but also effectively protecting the organization against cyber threats.

Can you name some of the distinct types of audits?

Some different types of audits are:

- **Information system (IS)** audits determine whether ISs and their related infrastructure are protected to maintain confidentiality, integrity, and availability, for example, to ensure that sufficient protection, privacy, or resiliency measures were taken to support the system.

- **Compliance audits** are used to determine whether specific regulatory requirements are being complied with, for example, ensuring that data acquired for one reason are not used for other reasons not consented to.

- **Financial audits** are used to determine the accuracy of financial reports, for example, to ensure that there are not any fraudulent use of government funds.

- **Operational audits** are used to determine the accuracy of internal control systems and help identify issues related to the efficiency of operational productivity within the organization, for example, to ensure that software is developed in line with company policies and procedures, while not producing unwanted defects.

- **Integrated audits** can be performed by internal or external auditors and are a blend of the other audit types used to assess the overall efficiency and compliance of an asset, for example, to ensure that banking loan approval systems are not discriminatory in their approval process.

- **Specialized audits** can include fraud audits, forensic audits, and third-party service audits, for example, detecting fraudulent credit card spending trends using company credit cards.

- **Computer forensic audits** are used to ensure compliance with the system during an investigation, for example, auditing the logs and activities of a computer system to determine the process that occurred in a computer system as part of a cyber incident.

- **Functional audits** are conducted prior to the implementation of new software to determine whether the software is functioning accurately, for example, to ensure that data loss protection software was sufficiently alerting or blocking attempts to extract sensitive information from a computer system.

What outlines the overall authority to conduct an IS audit?

The audit charter outlines the overall authority to conduct an IS audit and includes the auditor's objectives and responsibilities. Auditors' authority usually stems from the sponsoring stakeholder which is a senior leader within the organization.

What does the audit charter typically include?

The audit charter includes the scope, purpose, and objective of the audit team, the audit team's scope, the team members, and the responsibilities of each team member. The auditors can be an internal audit team, an external audit team hired by the organization, or an external audit team mandated by regulations or oversight requirements.

Describe the difference between a vulnerability and a threat

A **vulnerability** is a weakness in a system, which could be insecure code, weak security controls implemented, or a human factor. A **threat** is something that exploits this weakness, so this could be criminal hackers, ransomware or other malware, or something else, such as a hurricane.

Can you describe when you discovered a significant security vulnerability during an audit? How did you handle it?

This can be a tricky situation for an auditor, as a significant security vulnerability in a system does not always mean that this is a major security risk to an organization. At this point, it is important to document the particular finding in the system as well as any documentation of that vulnerability. Next, you should work with the organization's system administrator and security teams to raise the concern to them and inquire about any potential compensating controls that would thereby reduce the risk to the organization. Finally, you can document the leftover or residual risk to the organization.

For example, a manufacturing plant uses an old computer system that no longer receives any manufacturer patching or system security patches, has a long list of significant security vulnerabilities, and has not had any of the published patches installed. Updating systems or security patching on computer systems that run operational technology can negatively impact the machinery that is controlled by that computer system. Often, unless the machinery manufacturer has approved updates or patches, computer systems are not changed so they do not affect the machinery's functionality. To compensate for this vulnerability, operational technology is often segmented apart from other information systems so that the risks from those systems do not affect other information systems.

How do you ensure that your audit findings are communicated effectively to non-technical stakeholders?

Using the above example, an auditor can describe the lack of system and security patching on the computer systems that run the machinery as having the potential to risk the life and safety of plant employees if a threat actor were able to use the system vulnerabilities to remotely manipulate the equipment or cause it to fail. The auditor should also note that the compensating controls of segmenting the networking environment away from other information systems isolates it from other information systems and creates a gap in the systems, which would limit the actions of a malicious actor to one who might have physical access to the computer system in the plant.

How do you balance the need for thorough security audits with the practical constraints of time and resources?

As an auditor, you must balance the constraints, time, scope, objectives, and resources needed to accomplish the audit's goals. For example, when auditing a regional branch office of a bank, you might be tasked with ensuring that the bank is safely processing the banking transactions, deposits, and withdrawals, and safeguarding any associated records. While you may easily be able to audit the process flows and controls for protecting the records, doing a physical penetration test of the bank branch vault might not be a practical aspect of the branch audit.

Describe the term "assumed breach"

As technology systems (hardware/software) are created by humans and humans are subject to error, we can assume that a particular system might have some vulnerability. If we can all agree on that assumption, we should also all agree that there is the potential that your systems have already been breached. "Assumed breach" means you should be on a defensive footing and utilize a defense-in-depth approach and layering multiple layers of controls through the system.

This means that, as part of daily operations, you are ensuring that all secrets are protected, identities are verified to only be granted the needed access for the role you are in, and layering technical and administrative controls to protect organizational assets.

As part of this defensive posture, you are not just waiting to find **indicators of attack (IOAs)** displaying that you might be under attack from a threat actor; you should also be looking for potential **indicators of compromise (IOCs)**, which could be breadcrumbs that an attack or system compromise occurred unbeknownst to the system owners.

In this environment, there is little to zero inherited trust between people and systems. An example of an IOC is a workstation doing a callout to an unknown IP address, and an example of an IOA could be a PowerShell script being run on a user workstation where that user typically would not be running PowerShell scripts.

What is the difference between residual and inherent risk?

Inherent risk is the risk before any security controls are applied. This concept is often hard to understand for some, especially as there are so many controls that might come into play before a particular application or risk situation. It is important to talk through the nuances of this with your stakeholders before proceeding with the impending risk-based conversation.

Residual risk is the risk left over after applying security controls. Once you have arrived at this stage, it is important to consider the likelihood/frequency of the risk situations and the varying levels of potential impact when thinking about the residual risk and trying to quantify it.

What do you do if your client fails to see the risk in your audit report and recommendations?

The most important thing is to relate your audit findings to real-world examples that show how correcting the issue can benefit the team and organization.

For example, a manager of a department may ignore or not value your audit findings because they don't see the risk and think of correcting the issue as a burden on their already-reduced available time and budget.

Here, best would be in showing them how another departments implemented the changes that helped them reduce costs and improve their productivity.

Additionally, you should ensure you are providing the process you used to discover the findings and recommend fixing the issue. For example, if your audit discovers that employee user accounts were not properly terminated when the employees left, the client might just remove those accounts. This does not fix the problem for the client in the future when other employees are terminated. Instead, it is recommended that the client implements a process for identifying when employees leave the organization and how user account access is then removed.

When should you recommend the use of compensating security controls?

Compensating security controls are alternate security controls that organizations can use to fulfill a compliance standard, such as the **Payment Card Industry Data Security Standard** (**PCI DSS**) (`https://www.pcisecuritystandards.org/pci_security/`).

The alternate security controls must meet the intent and the same level of rigor as the original compliance requirement, provide an equivalent level of defense, and be comparable in the level of risk.

Compensating security controls are typically used when the organization has some type of constraint that prevents the implementation of the original security control in the compliance standard.

An example of compensating for security controls would be a small company that does not have enough staff in its financial department to have two or more people complete separate parts of a task. In this case, the small company might just use monitoring and analysis of logs and audit trails to track suspicious behavior in financial transactions.

What are some challenges when working with an environment based in the cloud or with a hybrid cloud/on-premises approach?

As a Security Auditor, your role is often to ensure that an organization has the right controls in place or that the controls in place are functioning in an effective manner as intended. Where controls tend to have a blurred line is when it comes to working in a third-party environment such as a CSP. When working with a CSP, there is often a reliance on them to implement certain controls, and you would inherit them from the CSP in your controls catalog.

For example, when working with a CSP, you are no longer responsible for things such as the physical safety, power requirements, and physical maintenance of the machines, so if you have controls like that for your physical environment, they would be inherited from the CSP.

Here is an example of the AWS shared responsibility model so that you can see which type of controls might have a customer or AWS responsibility:

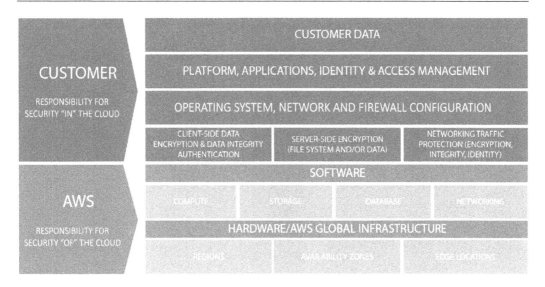

Figure 8.2: AWS shared responsibility model

Here is the Azure shared responsibility model:

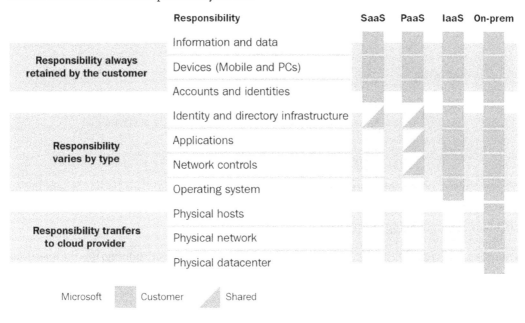

Figure 8.3: Microsoft Azure shared responsibility model

Here is an example of the shared responsibility model from **Google Cloud Platform** (**GCP**):

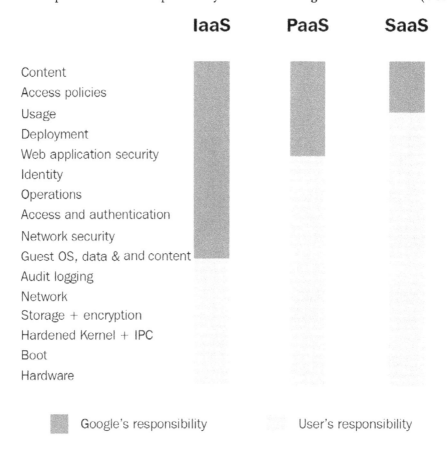

Figure 8.4: GCP shared responsibility model

As you can see, where the shared responsibility line is drawn is slightly different depending on the CSP, and it is your job to ensure that the organization is considering the right controls and that they are functioning effectively.

During an interview, you may experience a broad set of questions about auditing. Use the questions in this section as a guide and provide examples to the interviewer using your real-life experience with clients.

Summary

In this chapter, you learned what a Security Auditor is, the average salaries in the United States for a Security Auditor, and common questions you might be asked during an interview. Careers in auditing can be rewarding and lucrative, and since there is typically a shortage of auditors in many organizations, auditing can be a good career selection for someone who's new to IT and cybersecurity.

In the next chapter, you will learn some of the most common interview questions asked for a career as a malware analyst.

Join us on Discord!

Read this book alongside other users. Ask questions, provide solutions to other readers, and much more.

Scan the QR code or visit the link to join the community.

`https://packt.link/SecNet`

9

Malware Analyst

In this chapter, you will learn what a malware analyst is and the average salary range for this career in the US. You will also learn about the career progression options and some common interview questions for the role. This career is typically for someone who already has experience in another cybersecurity role, such as a SOC Analyst, and is not typically a career path for entry-level individuals. It's important to note that some of the questions listed in this chapter will require you to have background knowledge of assembly language. If you do not have this background knowledge, please be prepared to conduct additional research on the topics in the questions using educational materials (such as books, online courses, blog articles, and podcasts) that work best for your learning style.

The following topics will be covered in this chapter:

- What is a Malware Analyst?
- How much can you make in this career?
- What other careers can you pursue?
- Common interview questions for a Malware Analyst career

What is a Malware Analyst?

Malware Analysts analyze different types of malware to understand the threat they pose. This can include identifying the capabilities of the malware, how the malware functions, and identifying **indicators of compromise (IOCs)** that can be used to identify the malware. Other job titles you may see include malware reverse engineer, principal reverse engineer, and security researcher. A solid background in programming, networking, system administration, and operating systems is helpful for malware analyst roles.

Malware analysts also need to stay current on the latest threats. It is extremely helpful to have some knowledge of assembly language as you progress in your malware analyst career.

There are typically two types of malware analysts:

- **Escalation malware analysts**: They work with **incident response** (**IR**) teams and analyze malware samples to determine the functionality of the malware, what it might have done on the system, and artifacts (for example, URLs, filenames, and hashes) that will help the IR team look for the infection on other systems.

- **Collection malware analysts**: They usually work for a security company and go through public and private feeds to identify IOCs that can be ingested by security appliances to help the security company's customers identify and stop threats faster. As a collection malware analyst, you will typically specialize in a particular malware family and automate as much of the analysis process as possible for scalability.

The skills needed to be a malware analyst include curiosity, research skills, dynamic analysis (observing the malware behavior), static analysis (understanding properties of the file), and being able to articulate your process for analyzing a malware sample.

If you're looking to explore the malware analysis, there are some free resources available to help you get started:

- Malware Unicorn has a free introductory course on malware analysis: `https://malwareunicorn.org/workshops/re101.html#0`.

- Open Analysis Labs has a good YouTube channel for learning about malware analysis: `https://www.youtube.com/c/OALabs`.

- Here is a free Malware Analysis Bootcamp on YouTube from HackerSploit: `https://youtu.be/BjRMbe0-kLI`.

How much can you make in this career?

The salary for a malware analyst can range significantly from $76,500 to more than $165,000, depending on your location, the company you work for, and your experience.

What other careers can you pursue?

A career as a malware analyst can lead to more senior roles such as a principal reverse engineer, security researcher, senior incident responder, and even leadership roles like **SOC manager** or **chief information security officer (CISO)**. You can also use your malware analysis skills to move into full-time digital forensics roles and work with law enforcement agencies.

Common interview questions for a malware analyst career

The following questions include a focus on the analysis of **Portable Executable** (**PE**) headers and questions around assembly language. From my own experience in interviews, many questions I received for malware analyst positions were related to PE headers.

What is a PE file?

PE is a file format for executables, object code, and DLLs used in 32-bit and 64-bit versions of Windows operating systems. PE files can be loaded and executed across different versions of Windows. PE files contain essential information for the Windows operating system loader, such as headers, sections, and metadata that describe the file's structure, code, data, and resources. These files support various executable types, including EXE and DLL, and are integral to the Windows operating system's execution process. Malware analysts frequently examine PE files to understand how malicious software operates and to identify potential vulnerabilities or malicious behaviors within the executable.

Can you name the common headers in a PE file?

The headers are the DOS header (`struct_IMAGE_DOS_HEADER`), the NT header (`struct_IMAGE_NT_HEADER`), the file and optional headers that live within the NT header (`struct__IMAGE_FILE_HEADER` and `struct__IMAGE_OPTIONAL_HEADER` respectively), and the individual section headers (`struct_IMAGE_SECTION_HEADER`).

When opening an executable in a hex editor, the DOS header will occupy the first four rows (64 bits) in the hex editor and also include `MZ` in the magic number field. MZ stands for Mark Zbikowski, who was one of the original developers of the MS-DOS operating system. The MZ indicates that the file follows the DOS MZ format, which is a legacy part of the PE file structure.

The file header contains basic information about the file's layout and contains the following fields:

Starting Byte	Type	Information
1	WORD	Machine
3	WORD	Number of sections
5	DWORD	Time date stamp
9	DWORD	Pointer to symbol table
13	DWORD	Number of symbols
17	WORD	Size of optional header
19	WORD	Characteristics

Table 9.1: Header fields

The optional header contains the `magic` field, `AddressofEntryPoint`, the `BaseOfCode` and `BaseOfData` fields, the `ImageBase` field, the `SectionAlignment` and `FileAlignment` fields, the `SizeOfImage` field, and the `Subsystem` field.

The `magic` field identifies whether the executable is 32-bit or 64-bit. The `AddressOfEntryPoint` field is the address where the Windows loader begins the execution, and it contains the **relative virtual address (RVA)** of the **entry point (EP)** of the module. The `BaseofCode` field contains the RVA of the start of the code section, and the `BaseofData` field contains the RVA of the start of the data section. The `ImageBase` field is where the executable is mapped to a specific location in memory.

The `SectionAlignment` and `FileAlignment` fields are indicative of the alignment between the file and the memory locations. The `SizeOfImage` field contains the memory size that is occupied by the executable at runtime. The `Subsystem` field identifies the target subsystem for the executable file.

The individual section headers are contained in the section header table, which contains the `SizeOfRawData` field, the `VirtualSize` field, the `PointerToRawData` field, the `VirtualAddress` field, and the `Characteristics` field.

The `SizeOfRawData` field contains the information on the real size of the section within the executable. The `VirtualSize` field contains the size of the section in memory. The `PointerToRawData` field contains information about the offset where the section begins. The `VirtualAddress` field contains the RVA of each section in memory, and the `Characteristics` field displays the memory access rights for each section.

Section headers (`struct_IMAGE_SECTION_HEADER`) are used to define sections of the file, which can include information like the section name and the raw data size.

The most common section names are `.text`, `.data`, `.idata` or `.rdata`, `.reloc`, `.rsrc`, and `.debug`.

What is the difference between static and dynamic malware analysis?

In **static malware analysis**, you collect information about a malware sample without executing the malware to determine whether there are indicators present that the file might be malicious. This could include analyzing the hash of the file against malware signature databases, checking for suspicious strings such as suspicious domains and IP addresses, and identifying whether the file is packed.

Dynamic malware analysis allows you to analyze the behavior of the suspicious file to determine whether it is malicious. You should conduct this analysis in a sandbox environment. You can build your own sandbox environment using **virtual machines (VMs)** and there are also online sandboxes. **AnyRun** is one example of an online malware sandbox that executes your file and provides reporting on suspicious behavior, along with information about whether the sample matches any known malware. You can run a free scan on their website: `https://any.run/`.

As a malware analyst, you will typically use a hybrid approach for analyzing malware, where you begin with your static analysis and then analyze the behavior of the sample.

Walk me through your basic process for analyzing a malware sample

I start with static analysis and gather basic information, such as file hashes and metadata, and use tools like **PEiD** to identify the file type and possible packers. I then disassemble the code using a tool like **IDA Pro** or **Ghidra** to inspect the malware's structure, strings, imports, and exports, looking for indicators of malicious behavior.

I then proceed with dynamic analysis by executing the malware in a controlled, isolated environment such as a virtual machine or sandbox. I use tools like **Process Monitor** (Procmon) to observe files, registry, and process activities, and **Wireshark** to monitor network traffic. This helps me understand the malware's runtime behavior, including its communication patterns and any changes it makes to the system. Combining insights from static and dynamic analysis provides me with a comprehensive understanding of the malware's functionality and impact.

Can you explain what a sandbox environment is and how it is used in malware analysis?

A sandbox environment is a controlled, isolated virtual space where malware can be executed and analyzed safely without risking the host system. In malware analysis, sandboxes mimic an operating system to observe the behavior of malicious software in a contained setting.

Analysts use sandbox tools to monitor file modifications, network communications, process creations, and registry changes made by the malware to understand the malware's functionality and intent.

What are IOCs and how do you use them in your analysis of malware?

Indicators of compromise (**IOCs**) are pieces of forensic data that signify potential malicious activity on a system or network. These can include file hashes, IP addresses, domain names, URLs, registry keys, and unusual network traffic patterns.

In malware analysis, IOCs are used to detect and identify malware infections by matching these indicators against observed system behaviors and artifacts. During analysis, I would extract IOCs from the malware sample and use them to search for related malicious activity across systems and networks, enabling early detection and response. IOCs are often shared across the cybersecurity community to help other organizations protect against attacks and detect attacks faster. Many security tools ingest this type of data to provide automated responses to potential incidents.

What are some common signs of a malware infection on a system?

Signs of a malware infection can include performance issues on the system, random pop-up messages, unknown programs being installed, unauthorized changes to files or settings on the system, unknown processes or processes with similar names running, and antivirus software being disabled on the system.

What are YARA rules used for in malware detection?

YARA stands for **Yet Another Recursive Acronym**. YARA rules are used to identify and classify malware based on specific characteristics within files. YARA rules allow cybersecurity professionals to define strings, byte sequences, and other **Indicators of Compromise** (**IOCs**) that are typical of malware. YARA rules can be used to scan files, directories, or memory to detect malware by matching the malware to predefined patterns.

What are the common techniques used by malware to evade detection?

Malware can use techniques like code obfuscation, where the malware's code is deliberately made difficult to read and understand, and encryption, which hides the payload until it is executed. Polymorphic and metamorphic techniques are also used to alter the malware's code each time it infects a new system, making signature-based detection ineffective. Malware may also use rootkits to hide its presence by manipulating system processes and files.

Additionally, it might detect and evade virtualized environments or sandboxes commonly used by analysts. Anti-debugging measures can prevent or disrupt attempts to reverse-engineer the malware and the malware might be written to destroy the system if it detects an attempt at analysis.

What would you do if you found malware using obfuscation?

I would perform static analysis to identify the obfuscation methods being used (e.g., packing or custom encryption routines). Tools like PEiD can help me detect packers. Next, I would attempt to unpack using automated unpacking tools or manual methods. This could include debugging with tools like OllyDbg or x64dbg to look through the code and observe decryption routines. I would then perform dynamic malware analysis in a sandbox, which may help reveal the malware's payload for additional analysis.

What is fileless malware?

Fileless malware is a type of malware that is written to use legitimate processes and tools within Windows to avoid detection.

What is the significance of hash values in malware analysis?

Hash values are a unique digital fingerprint for files. This digital fingerprint allows analysts to quickly compare the hash values of suspicious files against known malware databases, speeding up the detection process. Hash values also help verify the integrity of files during forensic investigations, ensuring the files haven't been tampered with.

Walk me through how you would analyze network traffic for a malware attack

To analyze network traffic for signs of malware infection, I would start by capturing traffic using a tool like Wireshark. I then inspect the captured data for unusual patterns, such as unexpected spikes in traffic, connections to known malicious IP addresses, or communication on uncommon ports. Analyzing DNS requests can also reveal connections to suspicious domains, while inspecting packet contents can help me identify malicious payloads or command-and-control communication. Additionally, I look for indicators like repeated connection attempts, anomalous data exfiltration, or encrypted traffic that doesn't align with typical user behavior. By correlating these findings with known IOCs, I can detect and understand the nature of potential malware infections.

What is process injection?

Process injection is a technique used by threat actors to evade security tools and escalate privileges. Fundamentally, process injection is when the threat actor's code is run within an existing process.

This allows the threat actor access to the existing process' network resources and memory and possibly allows for escalated privileges for the malicious code.

You can read more about process injection on the MITRE ATT&CK website: `https://attack.mitre.org/techniques/T1055/`.

Name some tools you use when analyzing malware

This is subjective based on the malware analyst's answering, but some tools that can be used to analyze malware include the following:

- **Sysinternals**: `https://docs.microsoft.com/en-us/sysinternals/`
- **PeStudio** (Windows PE file static analysis tool): `https://www.winitor.com/`
- **Wireshark** (Network traffic analyzer): `https://www.wireshark.org/`
- **IDA Pro** (Disassembler): `https://www.hex-rays.com/products/ida/`
- **Flare VM**: `https://github.com/mandiant/flare-vm`
- **REMnux**: `https://remnux.org/`

What is Sysinternals and how is it used in malware analysis?

Sysinternals is a suite of system utilities built into Windows that is used in malware analysis to diagnose and troubleshoot issues related to system performance and behavior. Some of the tools include **Process Explorer**, **Autoruns**, and **Procmon** are used in malware analysis.

Process Explorer allows analysts to view information about running processes and their handles and DLLs. This helps analysts identify suspicious activity. Autoruns provides a view of all auto-starting locations and entries, which enables the identification of persistence mechanisms being used by malware. Procmon helps malware analysts capture real-time file system, registry, and process/thread activity, which offers insight into the behavior of the malware.

Explain how PeStudio is used in malware analysis

PeStudio is used to perform static analysis of PE files without executing them. It provides detailed insights into various data of the PE file, such as imported and exported functions, libraries used, and suspicious indicators like anomalies in headers or packed sections. PeStudio also checks for indicators of known malware by comparing against virus databases and examining the file's metadata, strings, and resources.

How is Wireshark used in malware analysis?

Wireshark is used to monitor and analyze the data packets being sent and received over a network. This can help identify suspicious activities in network traffic, like communication with **Command-and-Control (C2)** servers, data exfiltration, or connections to known malicious IP addresses. Examining packet contents, headers, and protocols help malware analysts trace the behavior of malware and collect IOCs.

Explain how IDA Pro is used in malware analysis

Interactive Disassembler Pro (IDA Pro) is a tool that's used in malware analysis for reverse engineering and dissecting executable files. It converts binary code into an assembly language format that analysts can examine to understand the internal workings of malware. IDA Pro provides a detailed view of the code structure, including functions, variables, and control flow, allowing analysts to identify malicious behavior, understand how the malware operates, and pinpoint specific routines or algorithms used by the malware.

Explain how Flare VM is used in malware analysis

The Flare VM is a Windows-based virtual machine for malware analysis and incident response. It offers a vast array of tools commonly used by malware analysts, such as debuggers, decompilers, disassemblers, and network analysis tools. Flare VM includes applications like **IDA Pro**, **OllyDbg**, **Ghidra**, and **Wireshark** for both static and dynamic analysis. Analysts use Flare VM to examine suspicious files, reverse-engineer malware, monitor network traffic, and perform forensic investigations.

Explain how REMnux is used in malware analysis

REMnux is a Linux-based distribution specifically designed for malware analysis and reverse engineering. It comes preloaded with a wide array of tools. These tools include static analysis tools like Binwalk and Radare2, dynamic analysis tools like Cuckoo, and network analysis tools like Wireshark and NetworkMiner. REMnux also supports the analysis of web-based threats and memory forensics.

Which header contains the field address of the EP?

The `optional header` contains the address of the EP.

What does the hex value in the optional header tell you?

The hex values in the optional header of a PE file provide important information about how the executable is loaded and executed by the operating system.

These values include the entry point address, image base, section alignment, size of the image and headers, subsystem type, and DLL characteristics. By analyzing these values, malware analysts can understand where the malware's execution begins, how it is mapped into memory, its intended execution environment, and any specific security mechanisms it might attempt to bypass. This helps you determine the malware's behavior and its potential impact on the system.

Which field shows you the total size of the header on the disk?

The `SizeOfHeader` field that is located within the `optional  header` field shows you this information.

Which field shows the number of sections in a PE file?

The `NumberOfSections` field shows this information and there are up to **65,535** possible sections.

How can you tell if a file is an executable file or a .dll file?

In the `Characteristics` field, the `IMAGE_FILE_EXECUTABLE_IMAGE` value will be set to 1 if it's an executable. If it's a .dll file, then the `IMAGE_FILE_DLL` field will have a value of 1.

You can learn more about image file headers from this Microsoft page: `https://docs.microsoft.com/en-us/windows/win32/api/winnt/ns-winnt-image_file_header`.

Describe the difference between the RVA and the AVA

The **Relative Virtual Address (RVA)** and the **Absolute Virtual Address (AVA)** are used to describe memory locations in a PE file, but they serve different purposes. RVA is the offset from the base address of the image when it is loaded into memory, representing locations relative to the image base without including the base address itself.

AVA, also known as **Virtual Address (VA)**, is the actual address in memory, calculated by adding the RVA to the base address of the loaded image. While the RVA indicates a position relative to the start of the image, the AVA provides the precise location in the process's address space, including the base address. Understanding both is important for accurate memory mapping and analysis during debugging and malware analysis.

Give me an example where you calculate the RVA

Calculating the RVA involves determining the offset of a function or data within a PE file relative to the image base.

For example, suppose the image base address is `0x00400000`, the VA of a target function is `0x00402000`, and the section's virtual address starts at `0x00002000` with a raw data offset of `0x00000400`.

First, calculate the offset within the section by subtracting the raw data offset from the file offset of the function. Then, add this offset to the section's virtual address to get the RVA.

If the file offset of the function is `0x00000500`, the RVA would be calculated as follows: RVA = `0x00002000` + (`0x00000500` - `0x00000400`) = `0x00002100`. That would make the RVA of the target function `0x00002100`. This calculation process helps to pinpoint specific locations within the executable for more detailed analysis.

What is the Import Address Table (IAT) used for?

The IAT is used in a PE file to dynamically link functions from external libraries, such as DLLs, at runtime. It contains pointers to the functions imported by the executable from these external modules, allowing the operating system to resolve and bind the addresses of these functions when the program is loaded into memory. In malware analysis, examining the IAT can reveal the external dependencies and functionalities the malware relies on, providing insights into its behavior and potential impact on the system.

What is the Import Names Table (INT) and how does it differ from the IAT?

The IAT points to the address of functions in memory and the INT points to the name of each function.

What are thread-local storage (TLS) callbacks?

TLS callbacks have been around for years and are an anti-analysis technique that allows malicious code to be executed prior to the `AddressOfEntry` point in the sample.

What are virtual tables (Vtables) in the C++ language?

For the classes that contain virtual functions, the compiler will create a Vtable. These Vtables house the entry for each of the virtual functions accessible by a class and store a pointer to the definition of the virtual function. The entries found in the Vtable can point to functions that are declared in the class or point to functions that were inherited from a base class.

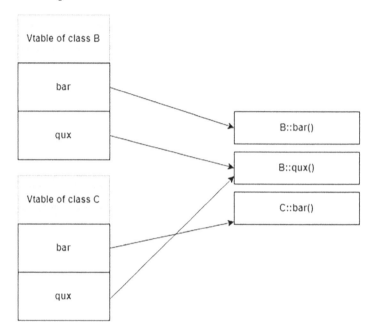

Figure 9.1: Example of a Vtable

What information does the .pdb file contain?

Program database (.pdb) files contain the debugging information for the program in Microsoft Windows.

How do disassembler programs know the names of functions and variables when you load an image?

The IMAGE_DEBUG_DIRECTORY contains the addresses of the raw data points that point to the functions and variables.

What is the difference between the VirtualSize and the SizeOfRawData fields?

When the image is on disk, SizeOfRawData is the size of the section. When the section is loaded into memory, VirtualSize is the total size of that section in memory.

What is the purpose of the .reloc section?

The relocation (`.reloc`) section contains information (variables and instructions) about where files should be mapped if they cannot be loaded using their preferred addresses that are defined with `ImageBase`.

You can learn more about relocation sections at `https://docs.microsoft.com/en-us/windows/win32/debug/pe-format`.

Can the .rsrc section be adjusted to provide executable permissions?

Any section can have permissions adjusted to allow for executable permissions. Permissions are defined in the `Characteristics` field. Typically, the `.rsrc` section contains arbitrary resources that the executable section of the PE can reference, such as bitmaps, icons, or other graphics. I have had interviewers ask about the `.rsrc` section because some malware authors embed malicious binaries within this section since the PE executable section references the `.rsrc` section.

If you are brand new to malware analysis, I recommend you pick up a copy of the *Practical Malware Analysis* book by Michael Sikorski and No Starch Press. I also recommend you check out the DFIR DIVA blog (`https://training.dfirdiva.com/`) as she posts many free resources for learning incident response, digital forensics, and malware analysis.

What are some tools that you can use to view PE header information?

The answer to this question depends on your favorite tools to use, but some are **PE Tools** (`https://petoolse.github.io/petools/`), **FileAlyzer** (`http://www.safer-networking.org/products/filealyzer/`), and **PEView** (`http://wjradburn.com/software/`).

In assembly language, what is the difference between MOV and LEA instructions?

The **load effective address** (**LEA**) loads a pointer to the item you are addressing. **MOV** accesses memory and loads the real value of the item you are addressing. LEA helps perform address calculations and store the result for later use.

You can learn more about the MOV and LEA instructions in the following resources:

- Assembly Tutorial: A closer look at the MOV instruction: `https://www.youtube.com/watch?v=eh3lnA_QykU`
- Machine Code in x86: `https://www.cs.uaf.edu/2016/fall/cs301/lecture/09_28_machinecode.html`
- Assembly MOV instruction: `https://stackoverflow.com/questions/24746123/assembly-mov-instruction`

- What's the purpose of the LEA instruction?: `https://stackoverflow.com/questions/1658294/whats-the-purpose-of-the-lea-instruction`
- Modern x64 Assembly 5: MOV and LEA: `https://www.youtube.com/watch?v=mKcWIA1vKOw`

Name some of the calling conventions in the assembly language

Some of the calling conventions are `cdecl` (C declaration), `stdcall` (standard call), Microsoft `_fastcall`, and Microsoft `_thiscall`.

Calling conventions are predefined protocols that determine how functions receive parameters from the caller and how they return a value.

Cdecl is a calling convention where the caller cleans up the stack, allowing variable argument functions. **Stdcall** is used primarily in Windows API functions, where the callee cleans up the stack, leading to more efficient function calls. **Microsoft _fastcall** optimizes speed by passing the first few arguments through registers instead of the stack, reducing overhead. **Microsoft _thiscall** is used for calling member functions in C++ classes, passing the this pointer in a register, ensuring that member functions can access class data.

How are standard calls and a C declaration different?

In standard calls (`stdcall`), the arguments are pushed from right to left and the cleaning of the stack is done by the `callee` function. In a C declaration (`cdecl`), the arguments of a function are pushed from left to right and the cleaning of the stack is done by the `caller` function.

What is the typical default calling convention in Windows C++ programs?

Standard call (`stdcall`) is the default calling convention you typically find in a C++ program, but you can always change the default calling convention in your compiler or IDE.

What is the output of an XOR operation if the operands are different?

The output using different operands would be 1.

Give an example of how to set the eax register value to zero.

xor eax, eax will add the content of eax to itself, which will set the value to 0. This is more efficient than using mov to change the value, as in mov eax, 0, and helps to reduce false dependencies.

Can you tell me the difference between a structure and a union in C programming?

A union only allows you to set the value of one variable at a time from the pool of variables. If you try to set the value of another variable, it will replace the value of the first variable since there is a single memory space allocated for the union.

A structure allows you to have multiple variables and you can set a value for each variable separately.

You have 4 GB of memory on a Windows system. How much is used for kernel space and how much is being used for user space?

In a 32-bit Windows system with 4 GB of memory, the address space is typically divided between kernel space and user space. By default, 2 GB of the address space is allocated to user space, where applications and user processes run, and the remaining 2 GB is allocated to kernel space, which is used by the operating system's core functions and drivers. This separation ensures that user applications operate in a protected environment, reducing the risk of system instability.

In some configurations, such as using the **/3GB** switch in the boot configuration, up to 3 GB can be allocated to user space, leaving 1 GB for kernel space, optimizing memory usage for applications requiring more addressable memory.

If you identify a high level of entropy in your analysis of a section, what could this indicate?

High entropy could indicate that the section is packed.

Cisco has a short blog post on entropy theory and how it can be used in malware analysis: `https://umbrella.cisco.com/blog/using-entropy-to-spot-the-malware-hiding-in-plain-sight`.

Can you tell me what the use-after-free vulnerability is?

Use-after-free vulnerabilities occur when freed memory is allocated to another pointer and the original pointer is then directed to newly allocated data. If this new data holds a class, then additional pointers may be spread within the heap data. If an attacker overwrites one of these pointers to point to their shellcode, then the attacker can execute arbitrary code.

Can you name some common process injection techniques?

Some of the common process injection techniques are DLL injection, PE injection, process hollowing, thread execution hijacking, hooking, registry modifications, **Asynchronous Procedure Calls (APC)** injection, **Extra Window Memory Injection (EWMI)**, Shims injection, and inline hooking.

These techniques are used by malware to execute malicious code within the address space of another process, often to evade detection and enhance persistence. **DLL injection** involves injecting a dynamic link library into a process's memory.

PE injection manipulates the PE structure to run malicious code. **Process hollowing** replaces the legitimate code of a process with malicious code after the process has been created but before it starts running. **Thread execution hijacking** redirects the execution of an existing thread to malicious code.

Hooking involves intercepting API calls to redirect them to malicious functions. **Registry modifications** can alter system settings to enable process injection. **APC injection** queues malicious code to run within the context of another process. **EWMI** leverages window properties to inject code. **Shims injection** exploits the Windows Application Compatibility framework to load malicious code. **Inline hooking** modifies the code of a running process to insert malicious instructions. These techniques allow malware to hide, persist, and execute without immediate detection by the security team.

Can you tell me what process hollowing is?

Process hollowing happens when it unmaps legitimate code from a process' memory and overwrites the space with malicious code.

This is a Black Hat conference presentation on process hollowing that can be helpful if you are not familiar with process hollowing: `https://youtu.be/9L9I1T5QDg4`.

What are some of the common Windows Registry locations that malware leverages for persistence?

They are:

`HKEY_CURRENT_USER\Software\Microsoft\Windows\CurrentVersion\Run`

`HKEY_CURRENT_USER\Software\Microsoft/Windows\CurrentVersion\RunOnce`

You can read about some of the persistence techniques used by APT groups on the MITRE ATT&CK website: `https://attack.mitre.org/techniques/T1547/001/`.

How many types of breakpoints are available in a debugger?

There are four types of breakpoints in a debugger. These are the software breakpoint, hardware breakpoint, memory breakpoint, and conditional breakpoint.

How does malware disable the Task Manager in Windows?

Malware can modify the security settings of the operating system by altering security policies stored at `HKEY_CURRENT_USER\Software\Microsoft\Windows\CurrentVersion\Policies\System`.

What are some indicators in a PE header that might reveal a sample is malicious?

If an executable is packed, you might see uncommon sections such as `.upx`, the `.rsrc` section might have processes without icons but with scripts or databases embedded within them, the `imports` of the section could be obstructed or missing, and the sections could have high entropy that could indicate malware.

Many intricacies of malware analysis could be asked during your interview. From experience in interviews, it's important for you as a malware analyst to be able to explain your process for analyzing a malware sample, including dynamic and static analysis, and be prepared to answer questions about PEs.

Summary

In this chapter, you learned about the malware analyst career and the salary range in the United States. You also learned how this can be a stepping stone into more advanced careers and learned some of the technical skills needed for the job.

In the next chapter, you will learn about a career as a cybersecurity manager, including common knowledge-based interview questions you might be asked.

Join us on Discord!

Read this book alongside other users. Ask questions, provide solutions to other readers, and much more.

Scan the QR code or visit the link to join the community.

`https://packt.link/SecNet`

10

Cybersecurity Manager

This chapter focuses on cybersecurity management roles. This type of role is a mid-to-senior-level role, typically requiring years of experience both in the subject matter as well as leadership.

The following topics will be covered in this chapter:

- What is a cybersecurity manager?
- How much can you make in this career?
- What other careers can you pursue?
- Certification considerations
- Example roles
- Common interview questions for a cybersecurity manager

What is a Cybersecurity Manager?

The role of a cybersecurity manager spans a broad range of types and responsibilities. At a high level, cybersecurity managers will either lead an entire cybersecurity program or a specific functional group that is part of a broader program strategy. Specifically, an organization might put a leader in place to lead its security operations program, which would include developing a strategy aligned with the desired security outcomes of the business from a tactical perspective. We will refer to this type of leader as a **Cybersecurity Program Manager**.

Within the overall security operations program strategy, there will be specific functional groups responsible for the day-to-day operations and execution of the strategy, such as threat intelligence teams, security analysts, and security engineers. These functional groups also require managers, and we'll refer to these managers as **Cybersecurity Team Managers** to prevent confusion.

Cybersecurity Program Manager

A **cybersecurity program manager** is an individual who effectively leads, communicates, and is responsible for an organization's cybersecurity programs. As part of a cybersecurity program's management, they are responsible for helping with the development of cybersecurity projects and infrastructure, as well as ensuring alignment with business programs and strategies. No matter how secure an organization may be, if the strategy introduces friction or roadblocks to daily operations and productivity, user adoption and executive buy-in will likely be an uphill battle.

Depending on the size and maturity of the organization, the cybersecurity program manager is usually heavily involved with the CISO and other members of leadership to ensure that programs align with their vision of how security should be handled. Some components of a cybersecurity program are compliance, governance, risk, security operations, and asset security (application, information, and infrastructure/endpoint security). These individuals are usually not involved at the operational level with all aspects of the program but more so at the strategic level, ensuring the program runs smoothly and on target.

Some of the certifications that can come in handy for an individual in this role are the following:

- **Certified Information Systems Security Professional (CISSP)**: `https://www.isc2.org/Certifications/CISSP`

- **Certified Information Security Manager (CISM)**: `https://www.isaca.org/credentialing/certifications`

- **Certified Information Systems Auditor (CISA)**: `https://www.isaca.org/credentialing/cisa`

- **CompTIA Security+**

- **SANS** (`https://www.sans.org/cyber-security-certifications/?msc=main-nav`) and **GIAC** (`https://www.giac.org/certifications/`) certifications

- **Project Management Professional (PMP)**: `https://www.pmi.org/certifications/project-management-pmp`

The considerations for certifications for cybersecurity team manager roles are very similar to those for cybersecurity program managers. I suggest evaluating the information and knowledge gained and the value they will add to your career when recommending certifications. Some values include bypassing HR filters to demonstrate to hiring managers that you have a level of knowledge in the field, or even just personal bragging rights while learning new information.

The types of cybersecurity programs may include, but are certainly not limited to, the following:

- Security Awareness and User Training

- Incident Response

- Security Governance and Risk Management

- Identity and Access Management

- Security Operations and Security Engineering

- Software Development Security

Cybersecurity Team Manager

A **cybersecurity Team Manager** is an individual responsible for a specific team within an organization's security program and any individuals who might be reporting to them. The size or maturity of the organization usually determines the team that a cybersecurity manager might be responsible for, and/or whether they would have responsibilities at the individual contributor level while being considered a process manager for delivering on the expected outcomes of processes like Reporting, IAM Account Reviews, IR Testing, etc.

There are several types of cybersecurity team managers; the first depends on whether the manager is a **technical manager** or a **people manager**. Depending on the size or maturity of the company and the type of product that they are delivering/creating, companies look for technical managers who have a deep understanding of technical requirements. These technical managers often grow directly from other technical roles into this type of leadership role, becoming responsible for managing different subspecialties.

The other type of manager in this category is the people manager who has a fundamental understanding of all the technical areas but does not go as deep as the cybersecurity specialists or consultants they might be managing. On the other hand, these managers can handle the required business relations, people, and processes. They can help teams remove some of the blocks and organizational processes that would otherwise slow or stop them from delivering results.

Cybersecurity manager roles

Cybersecurity manager roles and responsibilities vary, depending on the needs of the organization. As seen in the following descriptions, some organizations may choose to align managers to specialized areas of cybersecurity, such as critical infrastructure, applications, or the cloud:

- **Critical Infrastructure Security Manager**: Cybersecurity managers in this type of role are mostly responsible for ensuring the availability of the infrastructure by performing several due diligence activities, such as assessing and mitigating risk, monitoring for potential threats, and creating and testing incident response plans.

- **Network Security Manager**: Network security managers are typically focused on edge infrastructure and end user activity (firewall, IDS/IPS, web/content filtering, and so on) and are responsible for protecting corporate assets from external cyberattacks and insider threats. It is important that network security managers strike a balance between security and functionality, meaning that mitigating controls selected for implementation should not negatively impact employee productivity.

- **Application Security Manager**: Application security managers are tasked with leading teams of software developers to ensure secure coding teams by following best practices, such as dynamic and static analysis, input validation and output sanitation, proper encryption and authentication requirements, and access control.

- **Cloud Security Manager**: Migrating data to the public cloud introduces different data protection challenges compared to on-premises environments. One of the most common risks is data exposure resulting from a poorly defined cloud security strategy or misconfigured cloud security controls. Cloud security managers monitor cloud environments for vulnerabilities, threats, risks, and proper data access and cloud workload configuration controls.

On the other hand, cybersecurity managers could potentially be aligned to the functional areas of the cybersecurity program, such as security operations, the red team, and the blue team. The following list provides examples of cybersecurity manager roles:

- **Security Operations Center (SOC) manager**: The security operations team is a critical part of a cybersecurity program. At a high level, the team is responsible for the detection of and response to threats within the corporate environment, regardless of where they are discovered within the infrastructure.

  A sock manager is responsible for managing SOC analysts, in addition to defining policies, creating and refining security operations team processes, and working with security engineers to continue to grow the organization's capabilities.

- **Blue team/red team manager**: The blue team defends the enterprise. This team is responsible for monitoring and maintaining the company's security and network defense system against cyber threats, typically by working with (defending against) the company's red team.

  - The red team sits on the offensive side of the cybersecurity coin. This team helps the blue team by simulating cyberattacks designed to test the effectiveness of the security controls in place.

Individuals on these teams report to their team managers, who are responsible for working together to define the scope and rules of engagement for testing the organization's defenses.

Job titles and teams

There are several different job titles managers may hold according to the `https://www.cyberseek.org/pathway.html` website:

- Security Manager
- Information Systems Security Officer
- Information Security Manager
- Security Administrator
- Information Security Officer

The types of cybersecurity teams may include, but are not limited to, the following:

- Infrastructure Security
- Endpoint Security
- Cloud Security
- Storage Security
- Application Security
- Threat Intelligence
- Network Security Architecture
- Digital Forensics
- Defensive Analysts/Blue Team Analysts
- Offensive Analysts/Red Team Analysts

The following mind map may be helpful in understanding the different cybersecurity domains: `https://www.linkedin.com/pulse/cybersecurity-domain-map-ver-30-henry-jiang`.

How much can you make in this career?

Cybersecurity program managers can earn $90,000 to $150,000 based on their experience, location, and other factors. It could be higher, considering companies that might need program managers at this level include other bonuses and financial rewards in their compensation packages.

The salary range for cybersecurity team managers is extensive, as it runs from that of an individual contributor system or process owner, starting at $60,000, right up to the top end of the spectrum at $170,000 for those managing programs and who have one or more teams of stakeholders reporting to them. The location, specialization, experience, and area of responsibility will affect the salary earned in any specific position.

What other careers can you pursue?

As they progress in their careers, cybersecurity managers can anticipate roles that come with additional responsibilities beyond being individual contributors focused on the security of a single system. After managing multiple systems, managers can then begin leading teams. Continued growth usually leads to a manager becoming a director, who, in addition to being responsible for multiple managers and their direct reports, will also handle additional tasks, such as budgeting for the programming, road-mapping for the growth of the program, as well as interfacing and managing the relationships of the different leaders in the business.

With continued growth, here are some of the roles that you can progress to with continued focus and dedication:

- Chief Information Security Officer (CISO)
- Vice President of Security
- Senior Director of Cybersecurity
- Director of Information Security
- Head of Security Operations
- Senior Risk Manager
- Cybersecurity Strategy Consultant
- Cybersecurity Program Manager
- Senior Security Architect
- Principal Security Engineer
- Cybersecurity Compliance Director
- Head of Cyber Risk Management
- Cybersecurity Research Director
- Global Security Operations Director
- Senior Incident Response Manager

Common interview questions for a cybersecurity manager career

Here are some common questions that you may face during an interview for the position of cybersecurity manager.

What are the different types of programs you have previously been responsible for?

Rather than simply describing the technologies and systems you have handled, talk about the business problems, how specific technology was implemented to solve them, and the results achieved. For those responsible for multiple systems, discuss how the different systems were used/integrated/connected to help solve a more significant business problem and achieve results. If you are inheriting a legacy environment, discuss how you potentially optimized its use, maintained or replaced the system, and your results.

As technology evolves, managers must identify and replace outdated systems while ensuring that other processes are not disturbed.

How do you build business consensus on your security programs?

Building consensus and support for your security program starts with understanding the needs of the business and the way your program will help the company achieve its stated mission, vision, and goals. If the security program blocks and unnecessarily inhibits the actions of the business, it will circumvent the guardrails that the program is designed to deliver. Working with the business to create a pathway for success while minimizing any adverse risk and adequately informing the company of any risk and working with them to treat it is foundational to the program.

How do you help to drive cultural changes through your security programs?

Cultural change is about winning the hearts and minds of the consumers of the program and starts with C-suite and management buy-in to start integrating security and risk-based decisions into how they run the business. It all starts with awareness of the potential impact of security issues. Risk-based choices for a company could be the following:

- Risk-informed acceptance of a risk by the business
- Risk avoidance by avoiding risky business areas
- Risk mitigation by actively taking steps to reduce risk
- Risk transference by outsourcing aspects of risk or using insurance
- Risk ignorance by not acknowledging potential risks in an environment

- For more information on the risk management process, see the following: `https://www.pmi.org/learning/library/practical-risk-management-approach-8248`

Through security awareness, both at the management level and by encouraging awareness at the individual level, risks can be effectively managed. Having awareness at the individual level makes it personally applicable to an individual, rather than just part of their work role. Making the content or awareness training engaging rather than just informational will encourage more people in the organization to adopt a cultural change.

How would you demonstrate root cause analysis?

We would recommend answering this question as if you were already the cybersecurity manager for that company.

Cybersecurity managers hope to work with a team responsible for supporting the *incident response* to a situation, especially with a security implication. At the same time, the technology in question might not always be something that they manage directly. Some organizations conduct *postmortems* to better understand what happened during an event, what went right, what went wrong, what was the root cause (without assigning blame), and how things could be improved to avoid similar issues.

> *"Basic problem-solving. Demonstrate a methodical way of going from a symptom to a root cause and correction... It allows you to gauge [the] feedback mechanism (whether a system or stakeholders or both) to validate [the] plan of action."*
>
> —Omkhar Arasaratnam

How would you ensure your organization was prepared for an audit for SOC 2 compliance?

With any standard, process, or assessment, it is helpful to understand the standards against which compliance or success will be measured, and then to look inside your organization to see how close the standards are to being met. Using this gap assessment will guide the rest of the implementation. This approach can assess, address, and deploy many different compliance frameworks; in this example, we will focus on the **Service Organization Control 2 (SOC 2)** set of requirements.

SOC 2 is a set of compliance requirements and auditing processes targeted at third-party service providers that help companies understand external risks.

You are now the third-party vendor for another company, and they are looking at this report to help analyze the risk they will be assuming by partnering with your organization. It allows companies to determine whether their business partners and vendors can securely manage data and protect the interests and privacy of their clients.

SOC 2 was developed by the **American Institute of Certified Public Accountants (AICPA)** (`https://www.aicpa.org/`). Within its processes, there are two types of SOC 2 report:

- **SOC 2 Type 1** details the systems and controls you should have in place for security compliance. To prepare for this audit, you will need to provide auditors with evidence of these systems and controls and allow them to verify whether you meet the relevant *trust principles*. Think of this as a point-in-time verification of controls.

  However, as a good cybersecurity manager, you should strive to ensure that these controls continue to be maintained or grow in maturity over time. This will prepare you for the next type of assessment.

- **SOC 2 Type 2**: in this assessment, the auditor assesses how effective your program's processes are at providing the desired level of data security and management over time.

By planning to have validation of your controls continuously monitored, you can ensure that they function as designed. Additionally, you use the monitoring data to make improvements. You can then add additional automated reporting to encourage the growth and implementation of the controls.

Security compliance programs such as SOC 2 should just be a baseline for an organization.

The major focus of the SOC 2 compliance certification standard is listed as follows. To be successful, you should be able to demonstrate to the auditor that you meet the criteria that the AICPA has set:

- **Security**: The organization's system must have controls in place to safeguard against unauthorized physical and logical access. Be careful not to overlook controls for physical access and to understand the shared responsibility model when using cloud services.

- **Availability**: The system must be available for operation and must be used as agreed. Whether you're using cloud services or your systems, be sure to test failover capabilities and ensure that they function as intended.

- **Processing integrity**: System processing must be complete, accurate, well-timed, and authorized.

- **Confidentiality**: Information held by the organization that is classified as *confidential* by a user must be protected. Having a proper data definition and classification procedure from the beginning is extremely helpful compared to implementing one after a protection mechanism has already been deployed.

- **Privacy**: All personal information that the organization collects, uses, retains, and discloses must be in accordance with its privacy notice and principles. These are specified by the AICPA and the **Canadian Institute of Chartered Accountants (CICA)**. As global privacy standards expand, be sure to consider the implications they will have for your business models. For example, while you might not be doing business in Europe, you may still be subject to the General Data Protection Regulation.

SOC 2 is just one of many compliance standards that you might look to apply in your organization. The key is just to use them as a baseline for your security program, not a high watermark.

What are the key components of an effective cybersecurity strategy?

As you develop an effective cybersecurity strategy, you should consider aligning its business mission and the core ways it generates revenue. With that in mind, there should be a solid risk management and governance plan, controls that help mitigate environmental risk, continuous monitoring and logging, and a tested and validated disaster recovery and incident response.

How do I develop a cybersecurity strategy for my program?

The development of a cybersecurity strategy is, at its heart, an alignment with the business strategy. The first step is talking to business leaders and seeing where the business is heading or where it needs improvement, and then developing a strategy from that.

For example:

- If your salespeople inform you that SOC 2 is preventing them from working with larger clients, a solution for that could be included in the strategy, such as working with business leadership to assess the cost of getting the audit and maintaining compliance compared to the opportunity cost of any lost business.

- If your developers are looking to start bringing development in-house, you could help them with integrating security checkpoints during the SDLC process to find and discover bugs or coding issues before they become vulnerabilities.

Based on the desired future state of the business, you can conduct a gap assessment that will help you look at the work needed to complete objectives. A gap assessment allows you to compare the organization's current state with the desired future state and evaluate the changes required to get to the desired future state.

As you make any recommendations, start working with each line of business for as much alignment as possible. You can then work with them to integrate security initiatives within the business initiatives, which could help cover the implementation budget. Once alignment is achieved, ensure a buffer in the budget is included for dealing with unexpected incidents. In the case of budget cuts, ensure management is aware of functionalities or advances they will be sacrificing. To cover yourself, get decisions regarding budget cuts confirmed in writing/email in case they lead to a problem in the future.

How do cybersecurity program managers ensure alignment with business priorities?

As a cybersecurity program manager, you are responsible for ensuring that projects deployed within a cybersecurity program meet an organization's business needs. This organizational overview starts with clarifying the intent of the cybersecurity project and understanding the business problems they are looking to solve. The program manager must act as a business partner and influencer, delivering on their programs. They need to know what is happening within different lines of the business and the potential impact changes can have on the program.

While it is helpful for these individuals to be more technically minded so that they can look at the changing technological landscape, future technologies, and how they can help optimize their portfolio, it is not always necessary. Lack of technical understanding can be supplemented with the support of a team and/or outside consultants. Program management needs to be forward-thinking in helping businesses achieve their goals and have its finger on the pulse of current technology.

One of the early stages of this includes working with different lines of the business, understanding the impact of these changes (positive/negative), and then providing a top-level assessment before a project sponsor signs off on a project. At the more significant organizational level, changes to applications or services provided by the security department can have a knock-on effect on the workflow in different lines of the business, and this needs to be thoroughly understood and made a part of any risk-based decision before a project is approved or implemented. Starting with pilot programs can be a great way to test the effects of such changes within an organization and limit the results to a small subset.

How do cybersecurity program managers work to ensure that they can enable business programs?

The most critical aspect of a cybersecurity program manager's role is understanding a business application portfolio and ensuring that they can enable an optimized business mission by implementing security controls and procedures appropriately.

This means thinking ahead and remaining aligned with future business strategies and initiatives, ensuring the company's level of cyber-resilience improves (or, at a minimum, is maintained) as it continues to evolve.

For example, suppose your company is primarily on-premises in aging data centers and looking into ways of using **Cloud Service Providers (CSPs)** to help it migrate to a more scalable and resilient operating model. In that case, the cybersecurity program should help in that journey to enable secure options as the company shifts its business model. Whether it is considering **Platform as a Service (PaaS)**, **Infrastructure as a Service (IaaS)**, or **Software as a Service (SaaS)**, work with them to understand the shared responsibility models from each of the cloud providers so that they know their areas of responsibilities, the CSP's responsibilities, and where potential risks lie.

This responsibility includes understanding and working with the CISO and other security leaders on what changes they need to implement to optimize the business or update their security tools to meet changing needs.

How do you set up an application security program?

When working with developers or setting up an application security-focused security program, the concept of *shifting as far left as possible* is usually the first thing that comes to mind. When you are thinking about it from a people-process-technology perspective, starting with people is the first step you'll want to consider.

Let's say, for example, that developers are working with you to start addressing things such as preferred coding languages, their thoughts on security, and providing developer-focused security training. Supposing you have the data or statistics on the most common vulnerabilities discovered in their code in the past year, you can use them as examples with a member of the red team, showing how those vulnerabilities are exploitable. If you didn't have that data, you could start with the **OWSAP top 10 vulnerabilities** (https://owasp.org/Top10/). Providing developers with context and training on how to reduce those vulnerabilities and become security experts themselves expands the reach of the security department.

Combining processes and tools helps to complete a cycle by providing the right tools at the right time. Let's start with the **Integrated Development Environment (IDE)**; by providing tools that can help identify potential errors and vulnerabilities during coding and secure code training, you reduce the need for it to be reworked later as a bug or vulnerability. Next, having quick **Static Application Security Testing (SAST)** at the unit level ensures that the code doesn't have glaring vulnerabilities. As the code comes together, having **Dynamic Application Security Testing (DAST)**, **Runtime Application Self Protection (RASP)**, mobile code scanning, scanning containers, images, software repositories, and libraries helps make it more holistic.

It is beneficial to have a working group of development and security leaders deciding on the standards and implications, and changing the management approach to tackle applications that might have vulnerabilities and whether they have the approval to be deployed into an environment. There will be times when a business will need to deploy an application with known exposures to meet business needs. This working group can provide compensating controls for the environment and a business-as-usual remediation plan for any necessary software improvements.

How do you measure success in your security program?

The success of a security program should be measured based on the people, processes, and technology involved in the business. It could be focused on measuring the program's performance through reporting, focusing on the business's outcomes rather than just the number of IT events or logged items.

Another approach could be to measure the growth of the security program through the lens of the NIST Cybersecurity Framework or CIS Controls.

What approach might you take to help implement a safe business environment?

Start at the people level with organizational culture, which usually consists of the security awareness program. At the process level, the focus should be on business enablement, while at the technology level, it should be how security aids in securing the various levels of the organization's technology footprint.

While each company will have different metrics depending on its size, maturity, and industry, ultimately they should focus on how the security program has helped to mitigate or minimize risks for their company.

Helping the company's stakeholders implement security practices in their daily lives, whether that be at home or work, will help drive safe behaviors with technology. For example, providing users with password managers for personal use will make them feel more comfortable and minimize the harmful habit of password reuse. Many password managers include **Two-Factor Authentication (2FA)** features that allow them to display a rotating second-factor authentication token, helping to add a layer of security compared to simply using a username and password. Other aspects of assisting the people layer of an organization relate to helping them identify potential phishing emails and scams. Phishing and cons are increasingly arriving via social media and SMS, so helping employees be safe in their personal lives will also improve their behavior at work.

As mentioned previously, implementing secure password use and multi-factor authentication processes encourages the use of safe methods and mitigates risks around user identity takeovers and password reuse. Technology is also implemented as part of that process to help secure user identities. Some measurements you can look for are a reduction in account takeovers, a reduction in clicks on phishing emails, and an increase in the reporting of phishing emails.

This people-and-process approach should be used as a model for all other aspects of the business, and measuring success should minimize unsafe practices.

Improvements in vulnerability management programs should be made following a similar approach: understanding what can help improve processes and providing people with the necessary training and resources to improve outcomes. Measuring the ability of your team to mitigate known vulnerabilities within or before their assigned SLAs will demonstrate your program's ability to manage and reduce risks in this area.

How do you manage vulnerability management programs and the risk management involved?

At the program level, one of the many challenges an organization faces is ensuring that it has an excellent vulnerability management program, followed by program execution that leads to effectively managing the identified risks. There are a couple of foundational elements of vulnerability program management that need to be covered before discussing how we would manage the risk surrounding it for the organization.

The foundation of any vulnerability management starts with an asset management program. The first two controls of the CIS Top 18 controls (`https://www.cisecurity.org/controls/cis-controls-list`) demonstrate how important they are to an organization as they are at the top of the list. A good vulnerability management program needs a solid asset inventory management program that includes software, hardware, and understanding the implications of third-party software (having a software bill of materials helps to understand this: `https://www.ntia.gov/SBOM`) and services. This asset management program will let you know all the hardware, software, code, and other things within an organization's boundaries. Managers need to understand the current software or firmware levels of all these assets, what patches are available for the assets, and the potential vulnerabilities within that remain unpatched.

Sometimes, patches can break functionality or interoperability between assets. As software and hardware vendors release patches or updates to their software or hardware, it is crucial to understand the vulnerability that the patch or update is looking to mitigate and whether it might break any functionality or interoperability between other assets in an environment.

It is recommended that a patch be tested in a restricted sandbox environment to ensure that it works before rolling it out to the network.

Other aspects of vulnerability management that need to be considered are the prioritization of the application of the patches and the associated downtime, followed closely by understanding the resources of the various aspects of the business that would be responsible for the application process. Often, the resources responsible for the application of the patches are not the responsibility of the vulnerability management program. The program needs to develop an organizationally accepted **service-level agreement** (**SLA**) and work with resources to ensure that the patches are applied within the appropriate time frame.

Organizations need to develop a criticality rating (for example, critical, high, medium, and low) for risks and when those patches need to be remediated. This SLA is often created based on the criticality of the vulnerability for which the patch's risk is meant to be remediated. Some companies have developed a 7-, 30-, 60-, and 90-day approach for applying the appropriate patches. Unfortunately, that does not take into account the compensating controls within the organizational environment, so while a vulnerability might be rated as critical externally without controls, it might be different inside the environment.

Finally, now that you have identified, classified, and understood the risk implications for the organization, it's time to track them through remediation. Tracking vulnerabilities is the last critical component of the program, keeping leadership, developers, and stakeholders informed of the count, criticality, and remediation status.

How do you stay current with the latest cybersecurity threats and trends?

Keeping up with the latest cybersecurity threats and trends is a continuous part of the role, as often the organization looks to you to keep them informed. Some cybersecurity professionals keep an eye on Twitter for trends where security researchers share vulnerabilities and findings, sign up for alerts from security vendors, and subscribe to news alerts for vulnerability disclosures.

As a potential cybersecurity manager, during your interview, it is critical that you pull back from the interviewer's line of questioning and discuss these points:

- Where are the pain points of the security program and the business?
- What have they done to address them, and what have they tried that's failed?
- How can you help them solve their issues?

Summary

In this chapter, you learned what a cybersecurity manager is, their average salaries in the United States, certifications to consider, career path options, role types, and common questions you might be asked during an interview. Be sure to understand the difference between roles that require more people leadership than technical leadership and vice versa, as well as whether the role is more specialized or functional in nature.

In the next chapter, we will turn our attention to a role that is one of the industry's best-kept secrets: cybersecurity sales engineer.

Join us on Discord!

Read this book alongside other users. Ask questions, provide solutions to other readers, and much more.

Scan the QR code or visit the link to join the community.

`https://packt.link/SecNet`

11

Cybersecurity Sales Engineer

This chapter focuses on the role of the cybersecurity sales engineer, which is also commonly referred to as sales or solution consultant, solution architect, or simply presales. Although some organizations are moving away from using *engineer* in the title in favor of more business-focused titles, the sales engineer title is still very common and is therefore the title that will be used throughout the remainder of the chapter. When searching for jobs of this type, it would be wise to search all versions of the title to ensure the most complete search results.

The following topics will be covered in this chapter:

- What is a cybersecurity sales engineer?
- How much can you make in this career?
- What other careers can you pursue?
- What education and/or certifications should be considered?
- Common interview questions

What is a Cybersecurity Sales Engineer?

A cybersecurity sales engineer is a technical sales resource that supports the sale of cybersecurity technologies and/or services to businesses. This role is a hidden gem in the cybersecurity and broader tech industry; many technical resources shy away from this type of role mostly due to misconceptions regarding sales requirements. In most organizations, sales engineers are considered sales overlays, basically meaning they support the sales process but do not own it. This means that although sales engineers are often tied to the quota of the sales reps they support, they are not directly responsible for activities such as prospecting or cold calling, pipeline generation, and forecasting.

Sales engineers typically support one or more account managers (or sales representatives). While the account manager is responsible for managing the specifics of the sale (pricing negotiations, contract negotiations, term length, contract signature, and so on), the sales engineer is responsible for understanding the requirements of the business they are selling to and demonstrating how their products or services meet those requirements. These activities can include (but are certainly not limited to) providing a demo, delivering a **Proof of Value** (**PoV**) or **Proof of Concept** (**PoC**), documenting the scope of the project and implementation requirements, and articulating how their company's product or service is better than its competitors. Sales engineers have significant influence over a buyer's decision to select their product or service and are therefore a critical function of the broader sales team.

How much can you make in this career?

Before diving into specific dollar amounts, it is important to understand the pay structure for a cybersecurity sales engineer. It is not the pay structure most technical resources are used to, which, in my experience, is either a base salary or, in some cases, a base salary and a performance-based annual bonus. According to a recent industry study, the most common compensation structure for sales engineers, in general, is a base salary plus a variable component; the split is usually somewhere in the ballpark of 80/20 or 70/30, with the base salary being more heavily weighted. The base salary is a straightforward fixed amount, but the variable component will differ from company to company.

The variable component of the total compensation package is performance-driven. For sales representatives, the variable pay is usually commission-based, meaning they are paid based on how much they sell. The same is also true for sales engineers in many cases, but the variable pay is sometimes tied to **management by objectives** (**MBOs**). In this model, variable pay is based on the achievement of tasks assigned to the sales engineer in alignment with company goals. In rare instances, variable pay can be a combination of commission and MBOs.

A few other helpful terms to be aware of are listed here:

- **Quota**: The sales goal or target a seller must achieve over a specified period (for example, quarterly or annually)
- **On-Target Commission** (**OTC**): The amount of commission a seller can expect to earn if they reach their sales target
- **On-Target Earnings** (**OTE**): The sum of the base salary plus OTC

Let's walk through an example. A seller receives a compensation plan detailing the following: annual quota = $3,000,000, OTC = $100,000. This means if the seller achieves $3,000,000 in sales, they can expect to earn $100,000 in commission for the year. To take it a step further, if the seller's base salary is $100,000, their OTE is $200,000. Compensation plans can become incredibly complex when you get into the details of things like accelerators (which allow for overperformance) and overachievement bonuses, but we won't get into those.

It's important to note that technical sales is a high-risk, high-reward career when compared to traditional security roles. Referring to the previous example, $200,000 OTE may seem like an attractive offer, but keep in mind that $100,000 of that total is not guaranteed. If, for some reason, the seller is unable to close a single deal, they will only earn their base salary of $100,000. The upside, however, is that the seller could achieve more than $3,000,000 in sales, which would (or at least, should) result in more than $100,000 in commission.

As of June, 2024, the average base salary for an experienced sales engineer is in the $150,000–$175,000 range. From a total earnings perspective, if we take a base salary of $160,000 and assume an 80/20 split, the OTC would be $40,000, yielding OTE of $200,000.

What other careers can you pursue?

Sales engineers develop a unique blend of technical know-how, customer-facing skills, and business acumen. There are various career paths available based on these skills, such as sales management, which would require leading a team of sales engineers and/or account executives and driving revenue growth.

Product management is another viable path for a sales engineer (see *Chapter 12*). Sales engineers work closely with customers and prospects to understand their needs and pain points, which is similar to a product manager gathering needs and requirements for developing a new product or feature.

Consulting is another career path that may be of interest to a sales engineer. This role is also very similar to the day-to-day role of a sales engineer, as consultants typically leverage their technical and strategic expertise to advise organizations on their cybersecurity solutions and program strategies.

What education and/or certifications should be considered?

The path to becoming a cybersecurity sales engineer is not a linear one. Some individuals transition into the role from hands-on technical roles, while others come from business or educational backgrounds and learn the technical skills on the job. In general, a successful sales engineer will have a solid understanding of advanced security and/or security technology concepts and specialize in a specific product or service. They are technically minded, but also business savvy, and possess a balance of both the technical and soft skills necessary to build relationships, solve business problems with technology solutions, and simplify complex technical concepts. Sales experience is a definite advantage, in addition to experience in a customer-facing role.

Several organizations and individuals are working to bring more visibility to technical sales as a career. Within the technical sales community, a very small percentage of sales engineers are cybersecurity specialists, so there is no shortage of opportunities available to the right candidates. The following list highlights some resources worth exploring if a career in technical sales is of interest:

- **PreSales Collective** (`presalescollective.com`): PreSales Collective is a global community of presales (technical sales) professionals who connect to learn, grow, and move the profession forward. The organization provides access to webinars, blogs, podcasts, ebooks, and other resources aimed at helping members grow their careers as presales professionals. The *PreSales Academy* by PreSales Collective is a 10-week program that prepares individuals for a successful career in technical sales.

- *The 6 Habits of Highly Effective Sales Engineers* by *Chris White* is a book that helps sales engineers improve their sales skills. It is a great read for anyone interested in entering the field.

- *Mastering Technical Sales: The Sales Engineer's Handbook* helps individuals in technical sales navigate the technical sales environment and effectively support sales.

Given the consultative nature of the role, degrees and certifications in the sales engineer's area of expertise help establish credibility with buyers. Specific to cybersecurity, the following are certifications you could expect a senior sales engineer to possess (this is by no means intended to be an exhaustive list):

- **Certified Information Systems Security Professional (CISSP)** (`https://www.isc2.org/Certifications/CISSP`)

- **Certified Information Security Manager (CISM)** (`https://www.isaca.org/credentialing/cism`)

- **Certified Cloud Security Professional (CCSP)** (`https://www.isc2.org/Certifications/CcSP`)

- **GIAC Security Leadership Certification (GSLC)** (`https://www.giac.org/certifications/security-leadership-gslc/`)

- **Security+** (`https://www.comptia.org/certifications/security`) and **Systems Security Certified Practitioner (SSCP)** (`https://www.isc2.org/Certifications/sscp`) are also relevant for newcomers to cybersecurity who may not meet the experience requirements of some of the more advanced certifications listed previously. Often, training and certification specific to the product or service being sold will be required as part of the on-the-job training.

Common interview questions

The sales engineer interview experience can sometimes be long and demanding, but finding the right fit makes it all worthwhile in the end. The number of interviews can range from three to more than five. I once went through eight interviews, which included a presentation to a team of sales engineers, before receiving an offer. Eight interviews might sound like a lot, but sales engineers must integrate seamlessly into the teams they'll be working with and supporting. In a role where relationships can make or break a sale, cultural fit is as important as technical fit.

Throughout the process, you should expect to speak to any combination of sales reps and sales leaders, sales engineers and sales engineering leaders, a member of the human resources team, and maybe even members of the company's customer success and product organizations. Some interviews will be one-on-one, and others will be with a panel. The ultimate key to crushing the interview and landing this type of role is to be confident, relatable, charismatic, and memorable. Remember – people buy from people.

As previously mentioned, cybersecurity sales engineers require a mix of technical, interpersonal (or soft), and sales skills; you should expect questions that challenge your capabilities in each of these areas. Communication and presentation skills are at the top of the priority list in terms of skills a presales leader will want to validate, so you can expect some sort of presentation or pitch to be part of the interview process (a great storyteller always stands out when delivering a presentation). Depending on the company's product or service offering, technical requirements may be technology-focused (such as endpoint protection, firewalls, and so on) or outcome-focused (such as threat hunting or risk management services). In some cases, the role will require both.

I cannot stress to you enough the importance of what I'm about to tell you. Although this is a technical sales role, it is still a sales role. Your job will be to convince prospective buyers that your solution is best positioned to address their business needs, which means you will need to be convincing and believable. You must be a great listener, inspire confidence, and speak in the language and at the level of those you're speaking to. These tips are important to keep in mind during the interview process because the way you answer questions and articulate your value should be tailored to the interests of the interviewer. For example, you would want your technical skills to shine in an interview with a potential peer, but you may want your personality and sales or negotiating skills to shine instead in an interview with a sales rep or sales leader.

Selling is an emotional and psychological process. When you listen to a question, listen to absorb, listen to process. Don't make the dreaded mistake of simply listening to answer. Make sure you understand why the question is being asked so you can not only determine what to say but how to say it as well. Lastly, if you don't know the answer to a question, here are a few ways you can handle the situation:

- Request that the question be repeated. This gives you additional time to think and come up with your next move – *"Would you mind repeating the question for me? I want to make sure I fully understand."*

- Repeat the question. This also gives you additional time to process – *"Let me make sure I have this right. I want to ensure I'm answering the question the right way. Is what you're asking <insert your interpretation of the question>? And are you asking this because you're concerned about <insert your interpretation of the concern> or should I be thinking about this a different way?"*

- Throw a question back at them. This is one of my favorite tactics, especially when dealing with a difficult personality. Believe it or not, sometimes, a meeting attendee's sole purpose is to trip you up or invalidate your offering. I've had this happen to me in an interview. I handled it like a champ, of course, but it caught me a little off guard initially. Watch out for these types, and don't be afraid to challenge their way of thinking.

In some instances, I've even had questions be withdrawn after asking for clarification or additional context. I'll warn you, though, that this approach could backfire, so be careful. At the end of the day, you want to win over everyone you're engaged with, so you don't want anyone to feel like you're being combative.

In some cases, there's a bit of an art and finesse required to pull this one off – *"That's a great question. I'd be happy to walk you through that. But before I do, would you mind telling me if you're handling that situation in your environment today? It would be helpful for me to have that context as well as I demonstrate my approach."*

And if you think that even using the above tips to get more time will still not lead you to know the answer, then all you can do is:

- Simply say *"I don't know."* A common pitfall for sales engineers is believing they need to have all the answers. It's great to have most of the answers, for sure, but admitting you don't know something can boost your credibility and help establish trust. There's an art to this as well, of course. You never want to literally say *"I don't know."* Instead, say *"Hmmm… I don't think anyone's ever asked me that question before. I'll have to look into that and get back to you."*

The following is a list of interview questions that will help you prepare for a cybersecurity sales engineer interview categorized by type.

General/behavioral questions

Questions in this category generally assess the type of person you are and what your motivations are – basically, how are you wired? Cultural fit and personality are the primary focus here, but that doesn't mean you shouldn't take advantage of opportunities to put your other shining qualities on display as well.

Why are you interested in working with us?

I love this question. Why? Because it's the perfect opportunity to showcase so many things at once. In a very simple and concise way, you can describe a little about yourself, show that you've done your homework and researched the company, toss in a bit of flattery by describing what about the company sparked your interest, and put your sales and business skills on display by selling the interviewer's company right back to them. The hidden advantage is this question typically comes up pretty early in the conversation, so your answer could potentially set the tone for the remainder of the call.

Example: *"When I researched ABC (company), I was intrigued by the unique approach to mobile device management. The market is pretty saturated with vendors having similar offerings, but I see ABC's solution as a disruptive technology the industry has needed for a long time. This aligns well with my personality and motivations as someone who enjoys leveraging technology to solve complex business problems. I see a tremendous amount of opportunity to do exactly that with ABC."*

Why are you considering leaving your current role?

Be careful with this question. You'd be surprised at the number of candidates I've spoken with who have no problems bashing their current or previous employer. Maybe your boss drives you nuts, or perhaps you're bored or feel you're underpaid. These are all valid reasons for wanting to move on, but what value does that bring to a conversation during an interview? Not much.

Remember, the interviewer is asking questions to learn about you, not your previous employers. So, every chance you get, tell them more about you. And it wouldn't hurt to say something positive about your current role, either.

Example: *"I really am happy in my current role, and to be honest, I wasn't actively looking. But when the recruiter approached me about this role, I felt like I'd be doing myself a disservice if I didn't at least have the conversation, especially after researching the company a bit more. If I leave my current company, it will need to be for the right opportunity, and I feel like this could be it."*

What do you do for fun?

This is a simple question but can be challenging to answer if you're not prepared for it. When asking this question, the interviewer wants to know who you are outside the office. What are you passionate about? What are your hobbies? Are there any fun facts you can share? Family and pets tend to be pretty good conversation starters if you get stuck. Ultimately, you want to be relatable. And if you've done your homework on the interviewer, this is a great opportunity to touch on any shared interests you've uncovered.

What questions do you have for me?

When I conduct an interview, I learn most about a person based on the questions they ask – how they process information, what their priorities are, and, often, whether they're truly interested in the opportunity. Prepare your questions in advance and work with your recruiter or whoever your internal contact is to get a sense of whom you'll be talking to upfront; this way you can plan strategically to align your questions to the individuals you'll be speaking with. Ask questions that demonstrate your understanding of the business and your desire to be a valuable contributor to the organization.

Long interview processes can make this challenging. As a senior executive, I'm often one of the last people candidates speak with, and, a lot of times, they've emptied the tank by the time they get to me. If you find yourself out of questions about the company, ask questions about the person. Ask them what they enjoy about working at the company, or some of the challenges they face. What motivates them to come to work every day?

Lastly, don't forget the soft close. Ask the interviewer flat out if they have any concerns about your ability to fulfill the duties of the role. If the answer is no, good for you. But if the answer is yes, you can address the concerns on the spot. After you've nailed that part, ask about the next steps. It shows you're interested and gives you a sense of what's coming, timing, and so on, so you can prepare.

How do you handle a situation where a buyer isn't responding to your messaging the way you thought they would?

There are lots of hidden questions under the surface of this question. First off, you want to tell a story. If too many of your answers are theoretical instead of being based on a true story, you may come off as inexperienced. For example, don't say *"I would probably...,"* or *"I think I would...."* Provide a real-world example of a situation you handled. Talk about how it started out, how you recognized there was a disconnect, and what you did to get things back on track. You want to tell a story that ends with a win – even if it's more of a moral victory.

Who is the best manager you've worked for? What did you like about their management style?

This question has nothing to do with any of your former managers and everything to do with you. Your answer will indicate the type of environment you thrive in and where you might have some challenges. For example, if you say, *"My favorite manager was my old boss back when I worked at the pizza shop. We used to have so much fun hanging around the shop, playing video games and listening to music,"* well, it sounds fun, but doesn't say much about who you are or what you value in a leader or your work environment. Or maybe it does say something, but likely not what you want it to say.

On the other hand, if your answer is something like *"The best manager I ever had was back when I first started my career as a help desk analyst. It was a scary time for me as a newcomer to the team, but I remember my manager always making herself available to answer any questions I had and really caring about my success as a member of her team."* In short, you've said you value leaders with engaging and supportive leadership styles.

How do you overcome objections when working with a difficult buyer?

In sales, objections become a regular part of the job. Think about it: your job is to convince the buyer; the buyer's job is to pick the right product or service for the job. They will ask lots of questions and won't always agree with your answers or approach. The interviewer is asking this question to validate your ability to navigate a challenging situation. Make sure that your answer, which should be a real-life story, leaves them feeling confident in your ability to win over the buyer persona, sometimes referred to as a blocker.

What is your role in a sales opportunity?

Show that you're all in and fully support the deal end to end. Some sales engineers are more engaged in supporting deals than others. The more you show you care, the better.

Example: *"My role as the technical resource in the opportunity is to establish credibility and build trust with the appropriate stakeholders. Wherever possible, I build relationships that create additional paths to the win. My primary responsibility is to get the technical win, but I also support the opportunity through the sales win. If we don't get the sales win, the technical win doesn't matter as much."*

Interpersonal/communication skills

These questions usually aim to assess your ability to effectively communicate and build personal relationships. Be sure your answers reflect these skills. Some examples are provided here:

How would you approach explaining a complex concept to a non-technical audience?

When answering this question, it's important to demonstrate your ability to meet your audience at their level and speak their language. For example, your conversation style and talking points with a chief financial officer should be much different from those with a SOC Analyst. It's helpful to use analogies and tell stories that simplify the concepts and make them relevant to your listener. Again, if possible, tell a story about a time you've done this.

Example: *"When explaining a complex concept to a non-technical audience, I focus on simplifying the language, using analogies, and engaging the audience. For example, I once had to explain encryption to a group of business executives. Instead of diving into technical details, I said, "Think of encryption like sending a secret message in a locked box. Only the person with the key can open it." This analogy helped them grasp the concept immediately. To make it more engaging, I used a simple visual aid: a locked box and a key. I showed how a message gets "locked" and then "unlocked" only by the intended recipient. This visual demonstration, combined with encouraging questions, made the concept clear and memorable for everyone."*

Tell me about a time when you were able to convince a buyer to solve their problem in a different way.

The key here is to demonstrate your ability to influence a buyer without being confrontational. You don't want to appear as if you are too pushy or unable to meet people where they are. You basically want to replay the buyer's current concern and approach to them, then explain your recommended solution, highlight the benefits, and include success stories where appropriate.

Example: *"I remind myself going into every sales conversation that our buyers already have a picture of what they're looking for in mind and, most of the time, it looks nothing like what I have in my portfolio. This mentality prevents me from making assumptions and forces me to listen so I can visualize the buyer's desired outcome. In my experience, it's usually the process the buyer is highly opinionated about, so if I align with them on the desired outcome, I create the perception that we're in full agreement. This enables me to position my approach as a faster, more efficient path to the desired outcome instead of dismissing the buyer's thoughts on how things should be done. This approach has never failed me."*

How do you usually open your sales meetings?

In this question, the interviewer is interested in your approach to establishing rapport and building a connection with the prospective buyer. This could be through small talk or finding something that you both have in common to discuss.

Example: *"I like to open my sales calls with general conversation instead of getting down to business right away. Starting out this way encourages meeting attendees to let their guard down a bit and I'm often able to identify something we have in common so I can connect with them on a personal level."*

How do you decide what to include in your pitch or solution proposal?

The most important thing to articulate here is your understanding of what it takes to present an effective solution proposal. Sales engineer proposals should be comprehensive, persuasive, and aligned with the specific needs of the buyer. It is also important to strike a balance between including the right amount of technical details without including too much or overwhelming the buyer.

Example: *"It's important to me to include enough detail in my presentation to address all the concerns and requirements of the buyer, but not more detail than necessary. Too much information can be confusing and open the door to unnecessary questions. It's important that the solution proposal includes an introduction and overview of the buyer's needs, an overview of the solution to those needs – including technical details such as scope and implementation requirements – and the benefits and value of the solution. After I've prepared my presentation, I put it through what I like to call the "so what?" test. This means that anything in the presentation that doesn't have a specific purpose or add additional value gets removed."*

Sales skills

This area of the assessment is two-fold. On the one hand, the interviewer is seeking to validate your sales experience and your perspective on where your role as a sales engineer fits into the sales process. On the other hand, there is a very simple question – Would I buy from you?

Describe your current sales process and your level of involvement.

This question separates good sales engineers from great ones. There is an implied question behind the stated question, and that is: are you truly a sales-focused technical resource, or are you a technical resource that supports the sales team upon request? You'd be surprised, but many sales engineers don't love the idea of being considered part of the sales team, so their approach is to parachute in to support sales requests, then parachute out. Great sales engineers see themselves as part of the sales team and remain engaged throughout the sales cycle. When you answer this question, you want to demonstrate your knowledge of a typical sales cycle and how you are able to influence opportunities at the various stages.

Sales stages and definitions will vary from company to company, but a list of commonly used sales stages and high-level definitions is provided here:

- **Lead generation**: Find potential buyers.
- **Qualification**: Determine whether the buyer is a good fit for your product or service.
- **Discovery**: Understand buyer challenges, pain points, timelines, requirements, and so on.
- **Proposal/presentation**: Present the proposed solution and pricing.
- **Evaluation/negotiation**: Buyer determines which vendor will be awarded their business.
- **Win/loss**: Opportunity is closed as a win or a loss.

Sales engineers can be involved at any of these stages but are most involved in the middle of the sales cycle, meaning they typically enter during or after *discovery*, and, in most cases, will own the presentation of the proposed solution (the sales rep will typically handle pricing), and participate in the *evaluation* process (PoV, competitive comparison, implementation requirements, and so on). Sales engineers want to achieve the technical win, meaning the buyer believes their solution is the best fit, but sales reps own the final commercial and legal negotiations that ultimately lead to the overall sales win or loss.

What is your definition of sales success?

This is another question aimed at testing your sales IQ and level of buy-in as a member of the sales team. Articulate your understanding of the importance of the technical win, but also your awareness that the real success lies in the sales win. The sales win is how the sales team meets sales targets, retires quota, and earns commission – reiterate that you are a part of that team. Sales success should be rooted in financial goals, whether that's meeting quarterly or annual targets, driving $x\%$ in expansion opportunities, or growing your customer base by x number of new logos. At the end of the day, the goal is revenue generation, so make sure your answer aligns with that somehow. This is also a great opportunity to be seen as a team player.

How do you handle a situation where the solution you're selling may not be the best fit?

Hopefully, buyers who are not a good fit are disqualified during the *qualification* stage, which means a sales engineer would rarely encounter this situation. There are instances, however, where requirements are misunderstood, or a sales rep thinks there's a chance to win the business even though you may not agree. Remember, you're on the sales team, so your position should be to exhaust every option, including offering a solution that fits a subset of the requirements if you can't meet them all. Sales engineers are sometimes seen as a roadblock if they push too hard on requirements and introduce unnecessary friction. It's perfectly fine to stand your ground. You don't want to sell something that will lead to an unhappy customer, but you also don't want to appear inflexible during your interview. So, back to the main point, your answer should carry the theme of exhausting all options while maintaining the trust relationship and the integrity of the company. Who could be upset with that answer?

I'm a buyer who believes your solution is similar to a competitor's, and your solution is more expensive. Tell me why I should buy from you.

This is all about your ability to sell value. Don't make the mistake of focusing on features and functionality. Your buyer has clearly stated that their concern is price, which means you are very close to having the technical win – you just have to remind the buyer why you're better. To do this, remind them of why they began this journey in the first place. Remind them of the problem they're trying to solve and their desired outcomes, and why your solution gets them closer to that outcome, in a more effective, efficient, and sustainable way than the competition. Focus on the key differentiators, operational considerations, and competitor's shortcomings that support your solution. Hopefully, at this point in the sales cycle, you know who the competitor is; if not, ask. If the buyer declines to tell you, see if they'll at least tell you what they like about the solution so you can provide an apples-to-apples comparison and highlight the pros and cons.

Technical skills

Technical skill requirements will vary based on the product or service you're selling. For example, if you are interviewing with an endpoint security vendor, you can expect questions to be focused on things such as operating systems, ransomware, policy management, and so on. The following questions are more generic in nature and focus more on your understanding of the product or service, your ability to position it, and your perspective on continuous learning:

How much do you know about what we do?

Do your homework. Learn as much as you can about the company's offering. Become familiar with the value proposition, key differentiators, integrations, and top competitors: basically, whatever is publicly available to you. The way you answer this question shows the interviewer your level of interest, your ability to perform research and provide feedback, whether you understand the product or service offering, and ultimately, how steep your learning curve will be if you're hired.

Be sure to include details that demonstrate your understanding of the technical details associated with the solution as well. For example, your knowledge of compatible operating systems for an endpoint technology, or your awareness of the onboarding process for a services provider.

Are you familiar with any of our competitor's offerings?

Again, *do your homework.* Know who the key competitors are and how the company you're interviewing with stacks up against them. Be prepared to identify ways you'd win against those competitors in a sales opportunity.

How do you stay up to date with the latest trends, breaches, and emerging threats?

Reference any blogs, articles, newsletters, podcasts, and so on that you leverage to stay in the know. If you attend conferences and webinars, highlight those as well. Depending on how you're feeling about the conversation, you could even bring up a recent noteworthy event, your thoughts on emerging threats and trends, or simply something you found interesting during your research. (You can expect to find sales engineers at well-known security conferences such as **Black Hat** and **RSA**, typically answering questions, providing demos, and delivering presentations at their company's booth.)

How do you keep your technical skills up to date?

Whether the role requires hands-on skills or not, understanding how things work only makes you better at your job. If you have a home lab, great! If you don't, be prepared to discuss other ways you prevent your skills from becoming outdated – even if it's reading whitepapers and watching video tutorials. This is important because even if you're not part of the implementation team, your buyers and their teams will want to understand the integration process and requirements, where applicable. The more you know and feel comfortable with, the more confident you will be in your sales conversations.

By no means is the list of questions included here intended to be exhaustive. As you prepare for interviews with specific organizations, ask yourself how these types of questions might show up in your interview and how the interviewer might ask them. The more you're prepared to tie your experiences to the specific needs of the company you're interviewing with, the more you'll stand out as a candidate. Remember, in a role like this, every call is a sales call – including the interview.

Summary

A career as a cybersecurity sales engineer can be incredibly lucrative and rewarding. The key to success is always seeking to understand the reason behind the question or concern, providing conversational answers, telling stories (avoiding binary answers where possible), and making sure your audience remembers you. Have fun in your interview – be yourself. Sure, the company you're interviewing with is assessing whether you're a good fit, but you should be doing that as well.

Finally, I'll leave you with this golden nugget...

Turn your interview into a conversation. There are only so many questions that can be asked in a 30- to 60-minute window. The more time you spend having a conversation, the more the interviewer gets to know you and remember you as a person, not just another candidate. Also (and this is super top-secret stuff!), conversations help limit the number of questions the interviewer can ask. Now, go out there and crush it!

In the next chapter, we'll pivot to the role of a CISO. This is actually great context for a cybersecurity sales engineer because many of your conversations will be with CISOs and other cybersecurity managers.

Join us on Discord!

Read this book alongside other users. Ask questions, provide solutions to other readers, and much more.

Scan the QR code or visit the link to join the community.

`https://packt.link/SecNet`

12

Cybersecurity Product Manager

This chapter focuses on the role of the cybersecurity product manager. The role is also commonly referred to as product owner, **technical product manager** (TPM), or simply product manager. The primary role of the cybersecurity product manager is to collaborate with technical teams and business stakeholders to ensure that security products meet market demands and align with overall business strategies. The role requires a blend of technical expertise, market knowledge, and strategic vision to drive the success of cybersecurity solutions.

The following topics will be covered in this chapter:

- What is a cybersecurity product manager?
- How much can you make in this career?
- What other careers can you pursue?
- What resources, education, and/or certifications should be considered?
- Common interview questions for cybersecurity product manager

What is a Cybersecurity Product Manager?

The cybersecurity product manager is a strategic and technical resource responsible for guiding the development and success of cybersecurity technologies and services within a business. This role is often overlooked as a viable cybersecurity career path given that it is not a role you would typically find on a security team, but cybersecurity product managers are critical to the success of cybersecurity vendors and service providers. Many professionals assume being successful in this type of role requires deep technical skills alone, but this is not the case – it also requires strong strategic and business acumen.

More specifically, cybersecurity managers should possess a strong understanding of cybersecurity principles and technologies, in addition to strong communication, analytical, collaboration, and problem-solving skills. Although product managers are not typically required to have experience in project management (see *Chapter 13*), having the ability to lead projects and deliver products and features on time and within budget is critical to the success of this role. In most cases, product managers do not own the entire development process, which means they also need to be relationship builders to foster strong collaboration with development, marketing, sales, and support teams.

Product managers are responsible for keeping a pulse on market needs, defining product features, and ensuring products and services align with business objectives and customer expectations. Their most common activities include conducting market research, gathering and prioritizing product requirements, creating detailed specifications, and working closely with cross-functional teams to bring the product or service to market. Post-launch, they are responsible for tracking performance, gathering feedback, and continuously improving the offering. They are key influencers when it comes to defining the product vision, strategy, and roadmap.

How much can you make in this career?

As of June 2024, the average annual salary for a cybersecurity product manager in the US ranges from $80,000 for an entry-level product manager with 1-3 years of experience, to $180,000 for a senior-level product manager with more than 8-10 years of experience. Other compensation components often include bonuses and/or stock options. In addition to experience level, salary influencers for this type of role are location, specialization, and company size/type.

What other careers can you pursue?

A career as a cybersecurity product manager can open up a wide range of opportunities. The mix of strategic thinking, technical capabilities, business acumen, and leadership skills that product managers use daily makes the possibilities endless. On a more traditional path, a product manager could continue to climb the ladder to become a **Director of Product Management**, leading a team of product managers and overseeing a specific product or service portfolio, or even a **Chief Product Officer**, the executive responsible for setting and executing the overall strategy in alignment with the company's vision and goals.

If opportunities outside the world of product management are of interest, a **Chief Information Security Officer** role (see *Chapter 14*) would not be too farfetched for a more seasoned cybersecurity product manager. Beyond these paths, becoming a strategic consultant or starting a business in the tech or cybersecurity space are also viable options.

What resources, education, and/or certifications should be considered?

A successful career as a cybersecurity product manager requires a balance of technical skills, soft skills, an understanding of security concepts and challenges, and a strategic mindset. The following subsections highlight some resources worth exploring if a career in product management is of interest.

Education

While specific degrees and certifications are typically not a hard requirement for becoming a product manager, demonstrating your capabilities and experience will help differentiate you in the job market:

- **Bachelor's degree**: A degree in business, marketing, finance, strategy, computer science, information technology, cybersecurity, or a related field provides a solid foundation.

- **Master's degree**: For more senior roles, advanced degrees such as an MBA or a master's in cybersecurity can be beneficial.

- **Certified Information Systems Security Professional (CISSP)** (`https://www.isc2.org/Certifications/CISSP`)

- **Certified Information Security Manager (CISM)** (`https://www.isaca.org/credentialing/cism`)

- **Certified Cloud Security Professional (CCSP)** (`https://www.isc2.org/Certifications/CcSP`)

It is worth noting that there is no single certification body, authority, or standard for product management, so while product management certificates are not as useful in getting a job as you may be used to in the cybersecurity industry, the resources below can prove useful to get you started and provide more context to the role:

- **Certified Scrum Product Owner (CSPO)**

- **Pragmatic Institute** leverages real-world insights, actionable best practices, and tools in its training programs (`https://www.pragmaticinstitute.com/`)

- **Product School** provides training using real-world case studies and frameworks (`https://productschool.com/`)

Common interview questions for a cybersecurity product manager

Hiring managers will focus on the skills required to do the job, but they will also want to know you possess the qualities that separate a good product manager from a great one. Before you go into an interview, ask yourself if you possess the following qualities and how you plan to demonstrate them:

- **Initiative**: Are you able to drive project progress without the need for constant guidance or supervision?

- **Adaptability**: Product management involves what can sometimes feel like endless feedback and countless iterations. Is that something you can handle, or would you become impatient?

- **Decisiveness**: Are you comfortable making decisions when others are not, or when elements of the situation are uncertain?

- **Relationships**: Are you comfortable building relationships and collaborating with technical and non-technical stakeholders?

- **Prioritization**: Can you balance the different business priorities and requirements?

- **Product-market fit**: Are you passionate about solving user problems, evaluating their behaviors, and understanding their needs?

- **Analytical skills:** Can you drive high-quality products by collecting and assessing data and information?

General/behavioral questions

Tell me about a time when you had to adapt to a significant change at work.

I consider this a loaded question. On the surface, the purpose of this question is to see how open (or not) you are to change and how well you respond. When you add the product management lens, however, the role is all about change (making pivots, implementing feature requests, combatting competitive capabilities, and the list goes on and on). So, when you answer this question, make sure you highlight your ability to be decisive, flexible, and adaptable.

Example: *"In one of my previous roles, we were developing a cloud-based security platform. Three-quarters of the way through development, new data protection regulations were announced, requiring enhanced encryption and access controls. Faced with the choice between continuing our current plan or pivoting to meet compliance, I decided to re-prioritize development tasks and delay our beta launch by two months to ensure our product met the new standards. This decision, while challenging, protected our customers and strengthened our product's value."*

Describe a project you are particularly proud of and your role in its success.

Be careful with this question. If I am the hiring manager asking this question, the way you answer will tell me a lot about who you are and what you think of yourself. This is an opportunity for you to pat yourself on the back, and you certainly should, but your answer should demonstrate more than how awesome you think you are. Tie your accomplishments to the business outcome to really nail this question.

Example: *"I'm particularly proud of a project where I led the development of a next-generation firewall for a mid-sized enterprise. I coordinated cross-functional teams, gathered and prioritized customer feedback, and ensured the final product met the needs of the market. As a result, we launched the firewall on time, received excellent customer reviews, and saw a 30% increase in sales within the first quarter. My role in aligning the team and maintaining a clear vision was crucial to the project's success."*

How do you handle stress and tight timelines?

This is a large part of the job. Be honest and make this answer real – even if the response is not related to a specific project.

Example: *"This is easier said than done, of course, but I handle stress and tight deadlines by prioritizing, maintaining clear communication, and staying organized. I break down large tasks into manageable parts and set realistic milestones. I also find that regularly checking in with the relevant teams and stakeholders ensures we are all on the same page, stay on track, and address any issues promptly. From a mental health perspective, taking short breaks and practicing mindfulness also helps me stay focused and calm under pressure."*

How do you ensure continuous improvement in your product management practices?

Be specific when answering this question, as there are several things you'll want to highlight. Provide examples of actions and strategies that have been successful in previous roles and projects. Whether you focus on gathering feedback, staying up to date with industry trends and emerging technologies, or useful metrics and KPIs, tell stories that show you understand the importance of continuous improvement and are capable of this task.

Example: *"Continuous improvement is such an important part of a product manager's role. The approach that works for me is regularly seeking feedback from my team, stakeholders, and customers; typically, through surveys and feedback sessions. I also work extremely hard to stay up to date on industry trends by attending conferences, taking courses, and reading relevant publications. Retrospective analysis is important as well, so I conduct regular meetings with my team to reflect on our successes and challenges and identify opportunities for improvement. Embracing agile methodologies allows us to iterate quickly and adapt to changing requirements. Additionally, I implement KPIs and metrics to track our progress and measure the effectiveness of our initiatives."*

Interpersonal and leadership skills

How do you communicate complex technical information to non-technical stakeholders?

When answering this question, it's important to demonstrate your ability to communicate with stakeholders having varied levels of technical knowledge. For example, the way you communicate with an engineering resource should be much different from your communication style when providing updates to the non-technical executive team. For non-technical audiences, it's helpful to use analogies and tell stories that simplify the concepts. For more technically inclined individuals, it's important that you speak their language to ensure alignment. Again, where possible, tell a story about a time you've actually done this.

Describe your experience working with cross-functional teams. How do you ensure effective collaboration?

To answer this question effectively, provide a specific example of your experience working with cross-functional teams and detail your strategies for ensuring effective collaboration, including the context and challenges faced.

Example: *"In my previous role, I worked closely with engineering, marketing, and sales teams to launch several cybersecurity products. To ensure effective collaboration, I established regular meetings and utilized project management tools like Jira for tracking development tasks, Confluence for documentation and knowledge sharing, and Slack for real-time communication. I also promoted an open-door policy to encourage team members to share ideas and concerns freely. This collaborative approach helped us align our goals and deliver successful product launches on time."*

How do you manage cross-functional conflicts or conflicts within your team?

Here you'll want to emphasize open communication, transparency, and collaboration. You'll want to demonstrate that your approach is inclusive and that all parties have a voice in the discussion. Highlight the importance of remaining calm, de-escalating the situation, and reminding everyone that they are on the same team with a shared goal. The way you answer this question should demonstrate your ability to foster a positive environment and resolve conflicts effectively.

Below is a generic example, but if you can share a specific situation without implicating specific companies, teams, or individuals, or violating any laws or policies, it would add additional value to your answer.

Example answer:

"My approach is to focus on open communication, transparency, and collaboration. I address the issue directly by connecting with all parties involved to discuss their perspectives and concerns, understand the root cause of the conflict, and ensure everyone feels heard. I then remind everyone that we are all on the same team, working toward the same goal, and facilitate a constructive dialogue to identify common goals and find mutually acceptable solutions."

Give an example of how you motivated your team to achieve a challenging goal.

When answering this question, it is important to describe the specific strategies and tools you use. Gives examples that highlight the importance of a regular meeting cadence to keep everyone aligned and the use of project management tools for tracking progress and sharing knowledge. Mention how real-time communication platforms facilitate quick interactions and problem-solving. Emphasize promoting an open-door policy to encourage idea sharing and address concerns, demonstrating your commitment to fostering a collaborative and inclusive team environment.

Example: *"I recently led a project where the goal was to develop a new feature on an aggressive timeline. To motivate the team, I made sure they understood the importance and impact of the project on our company's success and our customers' security. I set up a transparent and achievable roadmap, with the project broken into manageable milestones. We recognized and celebrated small wins along the way to keep morale high and encouraged a collaborative environment where team members could share ideas and support each other. By providing regular feedback, removing obstacles, and maintaining an open line of communication, I ensured everyone stayed engaged and focused. As a result, we not only met the deadline but also exceeded our performance expectations, delivering a high-quality feature that received positive customer feedback."*

Strategic thinking and problem solving

How do you align your product roadmap with the company's overall business goals?

This question can make or break you. Your answer can put you in the pile of great, decent, or sub-par product managers. The question behind the question here is, how do you ensure YOU are aware or aligned with the company's goals because without that perspective, you're flying blind.

A great product manager ensures they are informed of the company's strategic goals and visions. This guidance should come from your leadership (i.e., the **Chief Product Officer** or **Director of Product Management**). When you answer this question, highlight your business acumen and how that can help the company that is interviewing you.

Example: *"To align the product roadmap with the company's overall business goals, a critical first step is gaining a thorough understanding of the company's strategic objectives and key priorities. I meet with leaders and key stakeholders to discuss the company's vision, market trends, and customer needs. I then categorize this information into product goals and initiatives and ensure all initiatives support broader business goals. I then prioritize based on the features and projects that drive the most value, balancing short-term wins with long-term strategic investments with stakeholders and continuously reviewing and adjusting the roadmap as needed. Lastly, to ensure our product development efforts are always aligned with and contributing to the company's success, I maintain regular communication with stakeholders and review and adjust the roadmap as needed."*

How do you evaluate the competitive landscape and differentiate your product?

This is another question aimed at testing your product management IQ. Articulate your understanding of the importance of competitive differentiation and how your approach helps the company maintain a competitive advantage. Make sure you emphasize your focus on market research and competitor analysis, then on understanding customer needs and identifying your product's unique value proposition. Feel free to use specific examples, such as conducting a strengths, weaknesses, opportunities, and threats (SWOT) analysis to compare capabilities.

Example: *"I start by conducting comprehensive market research, including competitor analysis, customer feedback, and industry trends. I use tools like SWOT analysis to identify our competitors' strengths, weaknesses, opportunities, and threats. By understanding the gaps and unmet needs in the market, I can pinpoint areas where our product can stand out. I also gather insights from sales and customer support teams to understand what customers value most.*

With this information, I prioritize features that offer unique value and address specific pain points, ensuring our product provides a compelling and distinct advantage over competitors. Regularly revisiting and updating this analysis ensures we stay ahead in a dynamic market."

How do you handle a situation where a critical vulnerability has been discovered in your product?

There are several qualities this question could assess based on your answer, such as communication and organizational skills. However, there is more value in demonstrating your ability to take a proactive and structured approach to resolving a difficult, high-impact issue. Describe how you would immediately assemble a cross-functional team, emphasize the importance of quickly identifying and understanding the scope and impact of the vulnerability, as well as quickly developing a mitigation strategy. Be sure to highlight the need for transparent communication, which could include, for example, conducting a thorough post-mortem to identify the root cause and implementing measures to prevent similar vulnerabilities in the future.

Example: *"When a critical vulnerability is discovered in our product, I immediately assemble a cross-functional team including engineering, security, and communication experts. We prioritize identifying and understanding the scope and impact of the vulnerability, and I ensure that the team works swiftly to develop and test a patch or mitigation strategy. Simultaneously, we communicate transparently with affected customers, providing them with timely updates and guidance on how to protect their systems. After resolving the issue, we conduct a thorough post-mortem to identify the root cause and implement measures to prevent similar vulnerabilities in the future. This approach ensures a rapid response, maintains customer trust, and improves our security processes."*

Describe a time when you had to manage conflicting priorities among stakeholders.

This answer should reflect your ability to gather information from multiple stakeholders, conduct a data-driven evaluation, and make balanced decisions. Start by explaining how you would facilitate a meeting with all stakeholders to understand their perspectives and concerns. Emphasize the importance of using data to evaluate the impact of each priority, such as assessing the consequences of technical debt versus the potential revenue from new features. Highlight how you create a balanced roadmap that allocates time for both critical improvements and high-impact features, demonstrating your ability to align teams on a common goal and effectively manage conflicting priorities.

Example: *"In a recent project, I faced conflicting priorities between the engineering team, who wanted to focus on technical debt, and the sales team, who pushed for new features to meet customer demands. To manage this, I facilitated a meeting with both teams to understand their perspectives and concerns. I then used data to evaluate the impact of technical debt versus the potential revenue from new features. By creating a balanced roadmap that allocated time for critical technical improvements while also prioritizing high-impact features, I was able to align the teams on a common goal."*

Describe your approach to managing risk in product development.

Start by explaining how you conduct an initial risk assessment to identify potential risks, including technical challenges, market uncertainties, and resource constraints. An effective way to differentiate yourself is to identify the distinct types of risk a product manager may need to address, such as value, usability, feasibility, or viability risk, and identify which type of risk your specific example addresses. Mention categorizing these risks by their likelihood and potential impact using frameworks like a **SWOT analysis** or a **Risk matrix**.

It would also be helpful to discuss your approach to mitigating the most significant risks, such as building contingency plans, allocating extra resources, or scheduling additional testing phases. Also, emphasize the importance of maintaining open communication with the team and stakeholders to monitor and reassess risks regularly, leveraging common approaches to risk management similar to the following steps:

- Identify risks
- Analyze the risk and measure the impact
- Prioritize risks
- Contingency planning
- Monitor and review

Example answer:

"I conduct a thorough risk assessment at the start of the project to identify potential risks, including technical challenges, market uncertainties, and resource constraints. I then categorize these risks by their likelihood and potential impact. Then, I develop mitigation strategies for the most significant risks, such as building contingency plans, allocating extra resources, or scheduling additional testing phases. Throughout the project, I maintain open communication with the team and stakeholders to monitor and reassess risks regularly. By staying proactive and adaptable, I can address issues promptly and minimize their impact on the project's success."

Technical knowledge

The types of technical questions you receive will vary based on the specific focus or domain of the product or service being developed. The questions below should give you a sense of what to expect.

What role does multi-factor authentication play in enhancing the security of a product?

When answering this question, start by defining what MFA is and explaining its basic principles. Discuss the three main types of authentication factors: something you know (e.g., a password or PIN), something you have (e.g., a smart card, mobile phone, or hardware token), and something you are (e.g., biometrics like fingerprints or facial recognition). Emphasize how MFA enhances security by adding layers of defense, which makes it harder for attackers to gain access even if one factor is compromised. Provide examples of common threats that can be mitigated by MFA, such as phishing and brute-force attempts. Provide a real-world example of how implanting MFA in a product led to notable security improvements.

Example answer:

"**Multi-Factor Authentication** (**MFA**) requires users to provide two or more verification factors before gaining access. These factors include something you know (e.g., a password), something you have (e.g., a mobile device or hardware token), and something you are (e.g., a fingerprint or facial recognition). By combining these factors, MFA adds layers of security, making it much more difficult for unauthorized users to gain access even if one factor is compromised. For example, if a password is stolen through a phishing attack, the attacker would still need the second factor, such as a one-time password (OTP) from a mobile device, to access the account.

In my previous role, we implemented MFA in our enterprise security software. Users had to enter a password and then verify their identity with an OTP generated by an authentication app on their mobile devices. This additional layer of security protected sensitive information and increased user confidence in our product's security measures. The implementation also reduced the number of successful unauthorized access attempts."

When developing a product that handles or accesses sensitive data, what techniques do you recommend for ensuring the confidentiality and integrity of data in transit and at rest?

A good way to start answering this question is by highlighting the significance of data integrity and confidentiality.

Be sure to discuss the importance of protecting data both in transit and at rest. Provide examples of specific techniques such as encryption, secure communication protocols, and access controls. Discuss the implementation of encryption standards like **AES** for data at rest and **TLS** for data in transit. Emphasize the importance of regular audits, monitoring, and using secure key management practices. In closing, provide an example from your experience of how implementing these techniques successfully protected sensitive data or improved a product.

Example: *"Ensuring the confidentiality and integrity of sensitive data (really, all data) is critical when developing a product that accesses or processes it. To protect data in transit, I recommend using secure communication protocols such as **Transport Layer Security (TLS)** to encrypt data as it travels between systems, preventing interception and tampering. For data at rest, employing strong encryption standards like **Advanced Encryption Standard (AES)** ensures that stored data remains secure and inaccessible to unauthorized users. Implementing strong access controls and conducting regular audits further protects data by ensuring that only authorized personnel can access sensitive information.*

*In my former role, I led the development of a financial application that required strict data protection measures. We used TLS to secure data transmitted between the application and our servers, and we encrypted all sensitive data at rest using **AES-256**, providing a strong layer of security in the event the storage media were compromised. We also implemented secure key management practices, ensuring that encryption keys were stored and accessed securely. Regular security audits and monitoring allowed us to detect and address potential vulnerabilities promptly. These measures collectively ensured the integrity and confidentiality of the sensitive data our application handled."*

How would you prioritize vulnerabilities discovered in a product you are developing during a security assessment, and what criteria would you use to determine the urgency of patches?

Explain the importance of vulnerability management in maintaining product security. Describe the criteria you use to assess the urgency of patches, such as the severity of the vulnerability, the potential impact on the system, exploitability, and the affected components' criticality. Highlight the use of standard frameworks like the **Common Vulnerability Scoring System (CVSS)** to quantify the severity. Discuss how you balance addressing high-risk vulnerabilities quickly while planning for less critical ones. Provide an example from your experience to illustrate your approach to prioritizing and patching vulnerabilities effectively.

Example answer:

Discovering and prioritizing vulnerabilities is critical to ensuring the security and reliability of a product. I use several criteria to determine the urgency of patches: the severity of the vulnerability, its potential impact on the system, exploitability, and the criticality of the affected components. Leveraging the **Common Vulnerability Scoring System** (**CVSS**) helps quantify the severity and provides a standardized way to evaluate and prioritize risks.

Recently, I led a project where a security assessment discovered multiple vulnerabilities. The first step we took was to prioritize and immediately patch the most critical vulnerabilities (high CVSS scores, easily exploitable, etc.) as these vulnerabilities would potentially have the most significant impact. We then scheduled patches for less critical vulnerabilities based on the importance of the vulnerable system and the potential impact of the breach. This approach allowed us to effectively address the vulnerabilities with minimal impact on the project timeline."

How do you ensure that a product complies with industry regulations and standards such as GDPR or NIST?

A good start to this question is to describe your approach to staying informed of any changes to current regulations and standards, such as **GDPR** or **NIST** – or even the introduction of new ones. Discuss your process of integrating compliance requirements into the development lifecycle. Lastly, emphasize the importance of keeping documentation up to date and keeping a pulse on any changes to requirements that may impact your products or current projects. Provide an example of how you've successfully achieved this in the past.

Example answer:

"To ensure product compliance, it is important to remain informed about current regulations such as GDPR and standards like NIST. This can be challenging, so my approach is regularly consulting legal and compliance teams, attending industry seminars, and reviewing regulatory updates. I integrate compliance requirements into the development lifecycle from the beginning, ensuring that all design and implementation phases consider these standards.

For example, in a project to develop a data protection solution, we ensured GDPR compliance by incorporating privacy-by-design principles. We conducted regular audits and involved our legal team to review our practices and ensure adherence to data protection requirements.

We documented all processes meticulously, from data encryption to user consent management, and implemented continuous monitoring to detect and address any compliance issues promptly. This proactive approach ensured that our product met all necessary regulations and standards, providing peace of mind to both our customers and stakeholders."

Summary

In this chapter, we learned about the incredibly important role of a cybersecurity product manager. Let's face it – without product managers, who would make sure the security tools we use in our environments solve the problems and address the threats we're worried about?

Join us on Discord!

Read this book alongside other users. Ask questions, provide solutions to other readers, and much more.

Scan the QR code or visit the link to join the community.

`https://packt.link/SecNet`

13

Cybersecurity Project Manager

This chapter focuses on the role of the Cybersecurity Project Manager. As we previously discussed, the product manager is focused on "why this product is important" and working to drive its implementation or maturation. A project manager will focus on the delivery and completion of a list of defined tasks and deliverables in order to successfully complete the project objectives.

The role is sometimes just called a **Project Manager** and it might be located within the **Project Management Office (PKMO)** of more mature organizations and outside of the specific **Information Security Office (ISO)**. Although some organizations are moving away from using Project Manager in the title in favor of more technically focused titles like **Technical Project Manager (TPM)**, the project manager title is still very common and is, therefore, the title that will be used throughout the remainder of the chapter. For Project Manager jobs of this type, it would be wise to search all versions of the title to ensure the most complete search results.

The following topics will be covered in this chapter:

- What is a Cybersecurity Project Manager?
- How much can you make in this career?
- What education and/or certifications should be considered?
- Common interview questions for the Cybersecurity Project Manager role

What is a Cybersecurity Project Manager?

A Cybersecurity Project Manager oversees projects that protect an organization's computer systems and networks from cyber threats, such as malware, phishing, and data breaches. Their role involves planning, executing, and closing projects related to cybersecurity measures.

They coordinate with IT and cybersecurity teams to ensure security strategies are effectively implemented and align with the organization's goals. This includes managing timelines, budgets, resources, and communication between stakeholders. They also assess risks, develop contingency plans, and keep up to date with the latest cybersecurity trends and threats to protect the organization's digital assets. Their goal is to maintain data confidentiality, integrity, and availability in a constantly evolving cyber threat landscape.

Cybersecurity project managers serve as vital links between the technical aspects of cybersecurity and a business's broader objectives. They work closely with senior management and stakeholders to ensure that cybersecurity initiatives align with and support the company's strategic goals. By understanding the business's vision and objectives, they tailor cybersecurity projects to bolster digital asset protection without hindering business growth. Their role involves meticulous planning and coordination to balance enhancing security measures with maintaining operational efficiency across various departments.

In managing risks, cybersecurity project managers run projects to identify potential threats and vulnerabilities and evaluate their impact on cybersecurity posture and business operations. They work on their projects with technical teams, translate complex technical risks into clear, understandable terms for non-technical stakeholders, and facilitate informed decision-making. This communication ensures projects comply with legal, regulatory, and industry standards, necessitating ongoing collaboration with legal and compliance departments. Regular reporting on project progress and outcomes helps demonstrate how security efforts contribute to regulatory compliance and the organization's security strategy.

Furthermore, cybersecurity project managers advocate for a strong security culture within the organization, emphasizing the importance of every employee's role in safeguarding digital assets. They lead educational initiatives on cybersecurity best practices, fostering a proactive security mindset across all company levels. This dual focus on aligning security projects with business objectives and promoting an organization-wide commitment to cybersecurity programs is a critical focus for cybersecurity project managers. They are crucial to the company's resilience and success in the face of evolving cyber threats.

How much can you make in this career?

Cybersecurity project managers are crucial in safeguarding organizations from cyber threats, balancing technical acumen with project leadership to protect digital assets. Their salaries reflect the importance of their role and the demand for their unique blend of skills. In the United States, the annual compensation for cybersecurity project managers typically ranges from $90,000 to upward of $150,000.

This variance is influenced by several factors, including the individual's experience, the complexity of the projects they manage, and the specific sector they work in. Notably, those in industries with heightened cybersecurity risks, such as finance and healthcare, may see salaries on the higher end of the spectrum due to the critical nature of their work.

The geographical location also plays a significant role in determining the salary of a Cybersecurity Project Manager. Regions with a higher cost of living or those recognized as tech hubs, such as the San Francisco Bay Area or New York City, often offer higher salaries to reflect the competitive market and elevated living expenses. Conversely, areas with lower living costs may provide salaries that, while lower on paper, still represent a strong purchasing power in the local economy.

For cybersecurity project managers looking to advance their careers and potentially increase their earnings, focusing on continuous professional development is essential for setting themselves apart in the job market. This can include staying abreast of the latest cybersecurity threats, technologies, and methodologies through self-study, workshops, webinars, and industry conferences. Such ongoing education ensures that they remain valuable assets to their organizations, capable of addressing the ever-evolving landscape of cyber threats with the most current knowledge and strategies.

Moreover, developing soft skills such as leadership, communication, and stakeholder management can further enhance a cybersecurity project manager's career prospects and salary potential. As they ascend to higher-level positions, their ability to effectively lead teams, manage complex stakeholder relationships, and contribute to strategic decision-making becomes increasingly critical. Specializing in high-demand areas of cybersecurity, even informally, can also make project managers indispensable to their organizations, positioning them for top-tier salaries and advanced roles within the cybersecurity hierarchy.

What education and/or certifications should be considered?

Becoming a Cybersecurity Project Manager typically requires a mix of formal education, practical experience, and specialized certifications. At the foundation, a bachelor's degree in information technology, computer science, cybersecurity, or a related field is essential, equipping candidates with crucial knowledge of computing and security principles. Advanced degrees, such as a master's in cybersecurity or IT management, can enhance prospects but are not always necessary. Practical experience is critical, with a pathway often starting in IT or cybersecurity roles such as systems administrator or network engineer. Transitioning into a project management role demands 3-5 years of direct cybersecurity experience, displaying an ability to oversee and implement security measures effectively.

Certifications can help demonstrate expertise, process knowledge, and competence in project management and cybersecurity. Some of the most prominent ones include **Project Management Professional (PMP)** and **Certified Information Systems Security Professional (CISSP)**. They help indicate proficiency in managing projects and deep technical knowledge in cybersecurity, respectively. In addition to certifications, it would also be helpful to have some hands on experience in leading or participating in projects, large and small, as each have their own challenges.

This combination of education, hands-on experience, and certification prepares individuals to effectively lead cybersecurity projects, safeguarding organizational digital assets against evolving threats.

Some of the potential certifications that could be beneficial for you in your career as a Cybersecurity Project Manager are:

- **Certified Information Systems Security Professional (CISSP)** (`https://www.isc2.org/Certifications/CISSP`).

- **Certified Information Security Manager (CISM)** (`https://www.isaca.org/credentialing/cism`).

- **Certified Cloud Security Professional (CCSP)** (`https://www.isc2.org/Certifications/CcSP`).

- **GIAC Security Leadership Certification (GSLC)** (`https://www.giac.org/certifications/security-leadership-gslc/`).

- **Security+** (`https://www.comptia.org/certifications/security`) and **Systems Security Certified Practitioner (SSCP)** (`https://www.isc2.org/Certifications/sscp`) are also relevant for newcomers to cybersecurity who may not meet the experience requirements of some of the more advanced certifications listed previously. Often, training and certification specific to the product or service being sold will be required as part of the on-the-job training.

- **Project Management Professional** (`https://www.pmi.org/certifications/project-management-pmp`), from Project Management Institute, helps with developing a structured approach for the delivery of cybersecurity projects.

Common interview questions for the Cybersecurity Project Manager role

When interviewing for a Cybersecurity Project Manager position, candidates must demonstrate a comprehensive blend of project management proficiency, cybersecurity expertise, adaptability, and leadership qualities. The interview process often delves into how candidates initiate, plan, execute, monitor, and close projects, specifically within cybersecurity. Interviewers aim to assess your familiarity with the project lifecycle applied to cybersecurity initiatives, seeking insights into your ability to manage these projects successfully under typical constraints such as limited time, budget, and resources. Providing examples from past experiences where you led cybersecurity projects to successful outcomes can vividly illustrate your capability and approach to project management.

Your knowledge of cybersecurity frameworks and standards, such as **NIST** or **ISO 27001**, is heavily evaluated as these are pivotal in guiding risk management and security measures. Interviewers expect you to be familiar with these frameworks and understand how to apply them effectively in different project contexts. In addition to understanding the fundamental controls, demonstrating how you stay abreast of the latest cybersecurity threats and technologies is just as important, as it highlights your commitment to continuous improvement. Discussing specific challenges you've encountered, such as a particular cybersecurity incident/event, and detailing how you handled these issues can demonstrate your problem-solving skills and depth of technical knowledge.

Risk management is another critical area of focus. You might be asked about your approach to conducting risk assessments or to describe a situation where you had to make tough decisions in the project to mitigate risk. These demonstrate your analytical skills and decision-making skills in minimizing potential security risks and understanding the impact on your projects. Additionally, as a leader on projects, you should also be able to work with other teams when vulnerabilities/risks are identified but might require outside assistance from other teams.

Some other questions might focus on your technical understanding from previous projects about security technologies and tools and how they've been implemented in projects. These questions are trying to gauge your hands-on expertise and ability to leverage people, processes and technology in securing organizational assets.

Finally, soft skills such as communication and leadership are paramount. You'll need to articulate how you translate complex technical risks into easily understandable terms for non-technical stakeholders, ensuring that the importance of cybersecurity measures is comprehensively understood across the organization. Your leadership style, particularly how it supports project success and team development in the context of cybersecurity, will also be a point of interest. Interviewers look for candidates who can lead by example, fostering a culture of continuous learning and adaptation, which are essential in responding effectively to the evolving landscape of cyber threats. Through a combination of project management skills, technical expertise, risk management strategies, and strong leadership, a Cybersecurity Project Manager can significantly contribute to an organization's resilience against cyber threats.

In Cybersecurity Project Manager position interviews, each question serves a dual purpose: it evaluates technical and project management expertise while providing insight into how candidates apply these skills in real-world situations. Here's how these questions unfold deeper layers of a candidate's profile.

Cybersecurity Project Manager questions

How do you initiate, plan, execute, monitor, and close a cybersecurity project?

This question tests a candidate's understanding of the project management lifecycle applied to cybersecurity. It reveals their approach to integrating security measures from project inception to completion, highlighting their strategic planning and execution skills.

We would recommend sharing this approach by sharing an example of your previous experience and how you accomplished this. Explain how you worked through the project phases and overcame any roadblocks.

More information about this process can be found here: `https://www.umb.edu/it/about/project-management-office/project-lifecycle-overview/`

Can you provide an example of a challenging cybersecurity project you managed? What was the outcome?

Interviewers look beyond theoretical knowledge to a candidate's practical experience and problem-solving abilities by asking for specific examples. It displays their capacity to navigate complexities and deliver tangible results, emphasizing the impact of their leadership on project outcomes.

We recommend considering specific project examples where you faced challenges, situations, and complications, and your participation in the resolution. Framing it specifically in this way, as you might not have implemented the resolution, highlights how your project management and issues resolution skills might have aided in its solution.

How do you manage project constraints such as scope, time, and budget in cybersecurity?

This question assesses a candidate's ability to work within every project's limitations. It's about understanding their resourcefulness in balancing these constraints while ensuring the project's cybersecurity objectives are met. Here, you might want to describe a situation in which you balanced the variables under your control within projects and worked with stakeholders on the tradeoffs when things changed during the project.

What frameworks or standards do you rely on for managing cybersecurity risks and why?

Knowledge of industry standards such as **NIST, ISO/IEC 27001**, or others is fundamental. This question evaluates a candidate's ability to apply these frameworks, ensuring the organization's cybersecurity practices are up to par with global standards. For example, we have previously recommended the NIST **Cybersecurity Framework (CSF)** as it acts like a Rosetta Stone translating the control requirements or overlaps from other frameworks. More information about how you can review the mappings to other frameworks can be found here: `https://www.nist.gov/ informative-references https://www.nist.gov/cyberframework`

How do you conduct a risk assessment for a new project?

Identifying, analyzing, and prioritizing risks is critical in cybersecurity. This question tests analytical skills and understanding of risk management principles, highlighting how candidates anticipate and mitigate potential threats.

Below is an example of a risk assessment/management cycle. While this process can be used to manage risk at a higher level for the company, it can also be scaled down to the project level.

Figure 13.1: Example of a Risk Assessment/Management Cycle Source: https://www.smartsheet. com/content/project-risk-assessment

Think about how you might have used a similar process as you were onboarding onto a new project to conduct a quick risk assessment of the current state of the project or readiness of the company to launch the project.

For more information, you can review the project risk assessment matrix by the Big Picture blog: `https://bigpicture.one/blog/project-risk-assessment-examples/`

What is a SWOT (Strengths, Weaknesses, Opportunities, Threats) analysis, and how can it be used in project risk assessments?

The SWOT analysis process can be used to assess the project resources (funding, support, people, technology, processes, etc.) and how they could help or hurt the likelihood of a successful project.

See an example of the SWOT analysis process below:

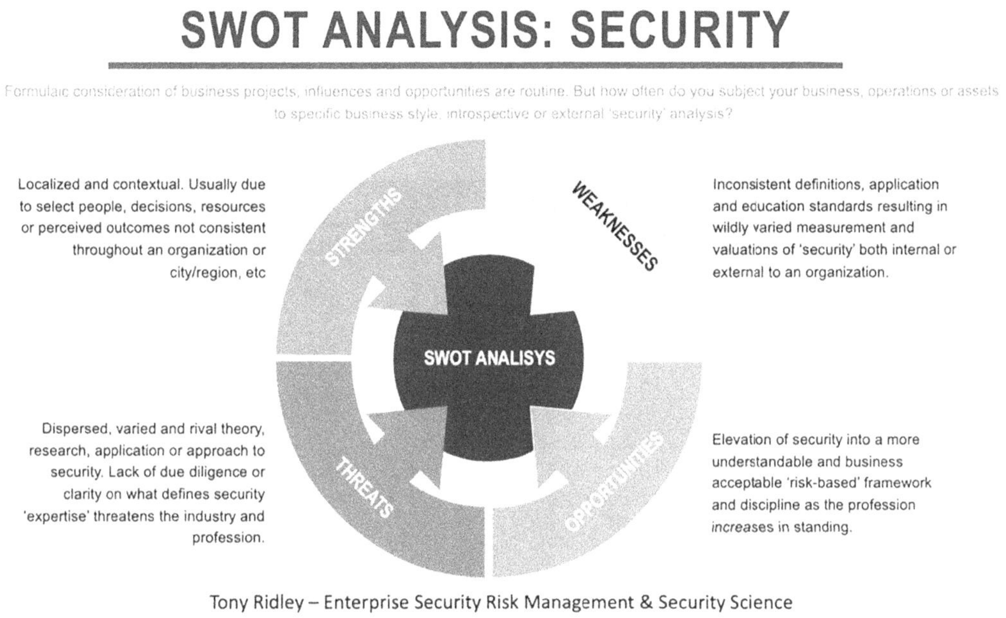

SWOT ANALYSIS: SECURITY

Formulaic consideration of business projects, influences and opportunities are routine. But how often do you subject your business, operations or assets to specific business style, introspective or external 'security' analysis?

STRENGTHS
Localized and contextual. Usually due to select people, decisions, resources or perceived outcomes not consistent throughout an organization or city/region, etc

WEAKNESSES
Inconsistent definitions, application and education standards resulting in wildly varied measurement and valuations of 'security' both internal or external to an organization.

SWOT ANALISYS

THREATS
Dispersed, varied and rival theory, research, application or approach to security. Lack of due diligence or clarity on what defines security 'expertise' threatens the industry and profession.

OPPORTUNITIES
Elevation of security into a more understandable and business acceptable 'risk-based' framework and discipline as the profession increases in standing.

Tony Ridley – Enterprise Security Risk Management & Security Science

Figure 13.2: Example SWOT Analysis
Source: https://www.linkedin.com/pulse/security-risk-management-swot-analysis-other-routine-tony/

Additional information about the SWOT analysis process in making business decisions can be found here: `https://www.rma.usda.gov/-/media/RMA/Publications/Risk-Management-Publications/swot_brochure.ashx?la=en`

How have you managed to balance your reporting and engagement responsibilities as a project manager?

As a project manager, you are responsible for engaging and keeping multiple stakeholders involved and informed on the project's status. However, not all stakeholders need the same level of information or engagement; you must know how to engage each stakeholder.

A helpful tool for understanding this is the **RACI matrix**, which stands for **R**esponsible, **A**ccountable, **C**onsulted, and **I**nformed. Using this matrix, you can map out how to engage the stakeholders on various parts or changes in the project.

As demonstrated in the matrix below, the initials from RACI are assigned to the various stakeholders and project components. It is critical to ensure that you document this at the start of the project so that you know who the critical stakeholders are and who you might just need to keep informed of the project status.

See an example of a RACI matrix below:

Define Initiation Phase	Project Manager	Project Sponsor	Project Analyst	Technical Analyst
Define Project Purpose	R	A	C	I
Define the Scope	R	A	C	I
Define Project Deliverables	A	C	R	C
Define Stakeholder Matrix	A	C	R	C
Define Governance Structure	R	A	C	I
Define the Implementation Approach	A	C	R	C
Define Risks and Issues	A	I	R	C
Project Charter	R	A	C	I

Figure 13.3: Example RACI Matrix
Source: https://www.navfac.navy.mil/Portals/68/NAVFAC/Careers/CC%20(WFD)/CCRC/CDC/
Developing%20Others/Infographics/RACI-Matrix.pdf

More information about the RACI matrix and lessons learned from using it can be found here: https://www.opa.hhs.gov/sites/default/files/2020-12/short-lesson-learned-raci-diagram-tool.pdf

How do you ensure your project team stays current with cybersecurity best practices and technologies?

This investigates a candidate's commitment to continuous learning and leadership in knowledge sharing. It highlights how they foster a culture of growth and adaptability within their team, ensuring the organization remains resilient against evolving cyber threats.

We recommend joining information security communities to help share information and best practices. Many of the credentialing bodies, such as **ISC2, IASAC, GIAC, CompTIA**, and others, have created communities around their bodies of knowledge and getting certification holders involved in sharing best practices and how they have handled different situations. Community members and companies also create communities on Discord, Slack, and other collaboration platforms to share knowledge and information.

General/behavioral questions

When interviewing for a Cybersecurity Project Manager position, organizations assess cultural, organizational, and team fit beyond the technical and project management skills. This assessment is crucial as it determines how well a candidate aligns with the company's values, work environment, and team dynamics. Here are some examples of general and behavioral questions that provide insights into these aspects.

What attracts you to our company's mission and the role of a Cybersecurity Project Manager here?

This question gauges a candidate's motivation and interest in the organization's goals and culture. It reveals whether the candidate has done their homework on the company's mission and values and how they see themselves contributing to those objectives through cybersecurity.

A personal tie to the business mission drives additional engagement and motivation to achieve positive project results. For example, as a Cybersecurity Project Manager for a children's hospital, you might want to drive positive impacts in children's medical research and cures so that they are available in case anything happens to the children in your family and community.

Can you describe a time when you had to adapt to a significant change in a project or organizational direction? How did you handle it?

Adaptability is a critical trait in the fast-paced world of cybersecurity. This question assesses a candidate's resilience and flexibility in facing changes, be it shifting project goals, adopting new technologies, or organizational restructuring. It highlights their ability to maintain effectiveness under evolving circumstances.

Give examples of how you might have dealt with shutting down or pivoting a project due to changes in the company focus. As a project manager, you should be able to assess the risks, positive and negative, of changing the project focus to stakeholders so that they understand any potential implications. This will allow stakeholders to make risk-informed decisions on the next steps for changing the project's course.

Tell us about a challenging team dynamic you've encountered. How did you navigate it?

Cybersecurity projects often require cross-functional collaboration. This question aims to understand how candidates work within diverse teams, manage conflicts, and foster a cooperative work environment.

It indicates their interpersonal skills and approach to building and maintaining productive team relationships.

These examples could include how you have dealt with stakeholders who might not have been on the same page about the course of the project or how to deal with specific roadblocks along the way. Project Managers often must coach their projects and ensure all stakeholders work well together.

What leadership style do you adopt when managing projects, and can you give an example of how this has led to a successful outcome?

This question explores the candidate's approach to leadership within the context of cybersecurity project management. It provides insights into how they motivate, guide, and support their team to achieve project goals, reflecting on their compatibility with the organization's leadership culture.

Sometimes, you need to be the visionary to show the stakeholders the potential future outcomes of a project or create harmony among the stakeholders to keep things running smoothly. Other styles include a servant leader who gets involved in driving the project's results, a coach who drives the stakeholders' growth while delivering project results, or a leader who sets high standards to follow or keeps the company/project in compliance with any applicable laws or regulations.

More information about different leadership styles can be found here, from the Project Management Institute: `https://www.pmi.org/learning/library/styles-leadership-avoid-mistakes-career-6280`

Margules, C. (2011). Styles of leadership—how to avoid "leisure suits," "high waters," and other career ending mistakes. Paper presented at PMI® Global Congress 2011—North America, Dallas, TX. Newtown Square, PA: Project Management Institute.

Describe a situation where you had to solve a difficult problem within a tight deadline. What was the outcome?

This question tests a candidate's analytical skills and creativity in crisis management. It reflects their ability to prioritize, take decisive actions, and leverage their team effectively under pressure, which is crucial for navigating the challenges inherent in cybersecurity projects.

An example of this is working with teams to set aside other tasks so that they focus on delivering the tasks for this project on time. Were you able to overcome the tight deadlines and deliver results on time, or was the project still late? The project's outcome is not always essential; your actions in overcoming it are crucial.

Describe your approach to dealing with stakeholders in a project where there were conflicting interests.

Effective communication and negotiation skills are crucial in cybersecurity project management. This question reveals how candidates navigate the challenges of aligning stakeholder expectations with project goals, highlighting their ability to mediate and communicate complex information.

Interpersonal/communication skills

Interpersonal and communication skills are critical for a Cybersecurity Project Manager. In my experience, the ability to effectively collaborate, and engage with your team and stakeholders, is crucial in aiding with the smooth execution of projects. Below are some targeted questions that help assess these skills, along with the rationale behind each.

How do you handle disagreements or conflicts within your project team?

This question probes a candidate's conflict resolution skills. Navigating disagreements and resolving conflicts constructively is crucial for maintaining team cohesion and productivity. It tests the candidate's emotional intelligence and approach to fostering a positive work environment, even under stress.

How do you communicate technical risks to non-technical stakeholders?

Translating complex cybersecurity jargon into understandable language for all stakeholders is invaluable. This question assesses a candidate's communication skills, which are crucial for ensuring informed decision-making across the organization.

For example, as a Cybersecurity Project Manager, I communicate technical risks by assessing stakeholders' technical understanding and using plain language to explain issues, such as describing a firewall as a security guard, GRC programs as guardrails on highways, or cybersecurity project managers as general contractors on a construction site.

Your role is to highlight the business impact, use visual aids like charts to illustrate risks, propose mitigation strategies, and maintain transparency about the risks' likelihood and severity. Your communication skills will also be critical as you must encourage questions to ensure stakeholders are comfortable and well informed.

Describe a time you had to present a complex cybersecurity concept to non-technical stakeholders. How did you ensure they understood?

The purpose here is to assess the candidate's ability to communicate technical information in an accessible manner. This skill ensures that non-technical stakeholders can make informed decisions based on the cybersecurity team's recommendations. It highlights the candidate's ability to tailor their communication style to their audience's level of understanding.

Can you give an example of when you had to persuade others to take a course of action they were initially resistant to?

This question evaluates a candidate's persuasive communication and negotiation skills. In cybersecurity, project managers often need to advocate for resources, changes, or specific security measures that are required immediately for the organization. The ability to persuade and gain buy-in is critical for successfully implementing cybersecurity initiatives.

This could be, for example, working with a collaborative team outside of your project to help you with resources to complete your project. However, it might mean they are delayed in their delivery objectives. One way to help you achieve this is to focus on combining the results teams achieve to help achieve results for the larger business mission.

Tell us about when you received critical feedback. How did you respond?

This inquiry sheds light on the candidate's receptiveness to feedback and their capacity for self-improvement. Effective communication is not only about conveying information; it is also about listening and adapting based on feedback. A positive response to criticism indicates a commitment to growth and a collaborative attitude.

An example is a senior leader sharing constructive criticism on how your delivery could have been improved or how you could have handled a situation better. How did you deal with this feedback? Did you incorporate it into your approach for future projects?

Describe how you've managed to keep a project on track despite communication challenges with a stakeholder or team member.

This question is designed to understand how the candidate navigates communication barriers or challenges. The candidate's approach to overcoming such obstacles and ensuring project continuity speaks volumes about their problem-solving skills and resilience.

For example, if you and a technical lead on a team disagreed on the approach to moving forward on overcoming an obstacle on a project. Did you come to a collaborative agreement on how to proceed, or did you let them proceed with their approach despite the potential risks and why? Being able to describe your actions will be crucial.

How do you ensure all team members feel valued and included in the project?

This question aims to understand the candidate's team dynamics and inclusivity approach. A project manager's ability to recognize and leverage each team member's strengths and contributions is critical to fostering an inclusive and motivated team environment. It also tests the candidate's awareness of diversity and their strategies for inclusive communication.

This question dives into your leadership style and how you engage your stakeholders. Sometimes, this means communicating more with some stakeholders so that they feel engaged or giving them more responsibility so that they can grow in their careers.

Technical skills

Assessing technical skills in cybersecurity project managers is essential to ensure they have the expertise to guide projects that protect organizational assets from cyber threats effectively. Here are questions focused on technical skills, along with the reasoning behind them.

What cybersecurity frameworks have you worked with, and how did you implement them in a project?

This question evaluates the candidate's familiarity with and application of cybersecurity frameworks such as **NIST**, **ISO/IEC 27001**, or **CIS Controls**. Understanding and implementing these frameworks is crucial for establishing and maintaining a comprehensive cybersecurity posture that aligns with industry best practices.

For example, if the organization is aligned with NIST controls, as the project manager, you would make sure that controls that align with that framework are followed and implemented as part of the project. If you are unfamiliar with the specifics of a particular framework, you must ensure that you collaborate with the organization's resources, which are familiar with the framework, and request external resources or training to help with the project delivery.

Can you describe when you had to assess and mitigate a cybersecurity risk in a project? What was the process and outcome?

The purpose here is to assess the candidate's risk management skills. A cybersecurity project manager's core responsibilities are identifying, assessing, and mitigating risks. This question provides insight into their analytical abilities, decision-making process, and effectiveness in safeguarding project assets against potential cyber threats.

Describe your experience with incident response planning and execution. How have you prepared your teams for potential cybersecurity incidents?

This question tests the candidate's readiness and strategic planning for cybersecurity incidents, a critical area in cybersecurity management. A proficient Cybersecurity Project Manager should have a solid plan for responding to incidents, minimizing damage, and recovering operations. Their answer can reveal their proactive approach and capability to handle crises effectively.

These could be focused on the project's risks and how, as project managers, they would deal with and overcome the obstacles surrounding organizations that have implemented best practices and comply with laws and regulations. The project manager does not need to be a technical expert in this area but should be aware of its role in the overall organizational risk strategy.

Explain how you evaluate and select cybersecurity tools and technologies for your projects.

This inquires into the candidate's process for researching, evaluating, and implementing cybersecurity tools and technologies. It tests their technical acumen, ability to align tools with project needs and organizational goals, and foresight in considering scalability, compatibility, and future-proofing investments.

Some cybersecurity project managers specialize in delivering projects in certain areas, such as **Identity and Access Management (IAM)**, **Security Operations Center (SOC)** modernization, or cloud modernization projects. By keeping up to date with the changing environment, tools, and modernization, you can grow your career and recommend projects that would help improve the organization.

Have you implemented any measures to ensure compliance with data protection laws and regulations in your projects? How did you approach this?

Compliance is critical to cybersecurity projects, especially given the variety and scale of global data protection laws and regulations. This question assesses the candidate's awareness and application of legal and regulatory requirements, approach to compliance within projects, and ability to navigate the complexities of legal compliance in tandem with cybersecurity objectives (See *Chapter 7, GRC/Privacy Analyst*).

Examples of some of these data, privacy, or regulatory concerns include **GDPR, CCPA, data sovereignty, PCI DSS, SOX compliance**, and other privacy laws specific to the company's location. See earlier chapters for more information on these topics.

Summary

The career of a Cybersecurity Project Manager is a dynamic and multifaceted journey that merges project management and cybersecurity disciplines to protect organizational assets from digital threats. This role demands a deep understanding of cybersecurity principles and technologies and the ability to apply project management methodologies to plan, execute, and oversee cybersecurity projects effectively.

Cybersecurity project managers are responsible for steering projects, implementing security measures, managing risks, and ensuring compliance with relevant laws and standards. They serve as the bridge between technical teams and business stakeholders, requiring technical understanding and strong communication, leadership, and negotiation skills to align project goals with business objectives and manage diverse teams.

The career path typically begins with a foundation in information technology or cybersecurity, often supported by a bachelor's degree in a related field. Practical experience in IT or cybersecurity roles provides the technical grounding necessary, while project management experience adds to the capability to lead and manage projects. Professional development through continuous learning and specialized certifications in project management and cybersecurity enriches their expertise and keeps them abreast of the latest threats, technologies, and best practices.

In their day-to-day work, cybersecurity project managers navigate a landscape of evolving cyber threats and rapidly changing technologies. They must be adaptable, continuously learning, and prepared to tackle challenges with strategic planning and effective team leadership.

Their success not only contributes to the secure operation of their organizations but also advances their careers, opening opportunities for higher-level management roles and specialization in cybersecurity areas of particular interest or demand.

This career is marked by a constant balance of technical problem-solving, strategic planning, and interpersonal interaction, offering a rewarding pathway for those passionate about protecting the digital frontier.

Join us on Discord!

Read this book alongside other users. Ask questions, provide solutions to other readers, and much more.

Scan the QR code or visit the link to join the community.

https://packt.link/SecNet

14

CISO

In this chapter, you will learn what a **Chief Information Security Officer** (**CISO**) is and the average salary range for this career in the US. Additionally, you will learn about the options for career progression and learn common interview questions for the role.

In this chapter, we will cover the following topics:

- What is a CISO?
- How much can you make in this career?
- What other careers can you pursue?
- Common interview questions for a CISO career

What is a CISO?

A CISO is a leader who is responsible for driving the information security program of an organization. Being a CISO is more than just understanding information security technologies and how the systems work together. It also involves understanding the business, including how the company operates, its regulatory and compliance environment, and its evolving external threat landscape.

As you can see, at this level in the organization, you require more than just technical expertise to be successful; you also need to have mastered working with the business and its different stakeholders, along with external parties, from vendors to regulators.

The core aspect of any CISO's role is to shape the company's information security program. However, before you can do that, you must understand the organization's people, processes, and technology, and pair that with leadership, culture, and strategy. You need to be able to use this strategy to inform, communicate, and coordinate how you are going to achieve shaping the security program across the different parts of the business.

How much can you make in this career?

The CISO compensation range greatly depends on the locality of the role, industry, and organization size. I have seen salaries ranging from $150,000 to $2.5+ million in total compensation, with salaries averaging between $350,000 and $450,000. At this stage in your career, salaries start to separate into multiple buckets to include base compensation, bonuses, restricted stock units, company stock, and other items designed to tie compensation to the company's performance. Another form of payment is *vested* rewards such as stock or stock options, which are granted based on longevity with companies based or operating in the US.

What other careers can you pursue?

This role is often looked at as the pinnacle of cybersecurity leadership roles, as individuals in this role might be operating with and be peers with individuals in other CxO roles.

A career as a CISO means that you have mastered the ability to understand the business, its risk drivers, and its vulnerabilities, and you can design mitigation programs around it. CISOs have a broad offering of careers to consider, including board advisors, other CxO roles, consultancies, **vCISOs** (either **virtual or fractional CISOs**), and more. Sometimes, CISOs will leave their positions at mature companies and go back to start-ups as head of information security due to the potential upside of joining an early-stage start-up before it becomes public or acquired.

Common interview questions for a CISO interview

The following is a list of interview questions that could prove useful in preparing for a CISO interview:

Why are you in cybersecurity?

While this might seem like a simple question, part of being at the leadership level is being able to easily communicate with stakeholders and other leaders who might not understand how these terms relate to them in the organization.

The approach we would recommend is sharing your "why" for cybersecurity. This could be a personal story of something that drew you into computers, security, business information systems, audit, etc. This personal story often helps share why you care so much about cybersecurity, and why you would be focused on ensuring that the business operates in a safe manner while delivering on its business mission.

I would recommend including different aspects from the overview of cybersecurity below in your personal story to draw out the correlations to cybersecurity and why you would be a great leader in this role.

Cybersecurity is in the space of protecting organization assets (such as data, processes, and infrastructures) in digital form from unintended actions that affect the confidentiality, information, and integrity of those assets. It is often considered a subset of information security, which also includes the protection of information in analog form.

How would you describe this concept to organization leaders who might not understand what is or isn't a risk?

At this level, it is more about being able to communicate the concept of cybersecurity to others. So, while you might have had a long career focused on getting into the technical details of risk, let us ensure that we can still clearly and concisely communicate this to other leaders.

A threat is something that has the potential to negatively impact an organization. These can range from a variety of different sources, including natural disasters (such as hurricanes, earthquakes, and floods), threat actors (an individual or group that wants to impose damage or harm on a person or organization), or other categories such as economic and regulator impacts on an organization. A risk is a function of that threat, exposure to that threat, the likelihood of occurrence, and any response or lack of response to the threat.

How do you measure risk as a CISO?

First, to be able to measure risk, you must be able to define it. Risk is the potential for something to affect an organization or business. This can be both positive and negative. Usually, risk is defined as a function of the potential impact of an adverse event, times the likelihood of that event. *Ryan Leirvik* mentioned in his book *Understand, Manage, and Measure Cyber Risk* that technology is inherently flawed because it is created by humans, and humans create technology.

I would like to calculate risk from another field of study, as it is more inclusive of all the elements that we need to think about:

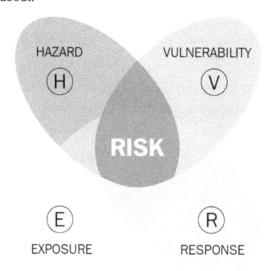

Figure 11.1 – Risk and its relations to its variables

A **vulnerability** in technology is the first element of risk that a CISO will measure. One of the ways to measure the size of a vulnerability is by using the **Common Vulnerability Scoring System (CVSS)** score (`https://nvd.nist.gov/vuln-metrics/cvss`). This uses **Common Vulnerabilities and Exposures (CVE)**, which is a list of publicly disclosed computer security flaws. The CVSS score uses several aspects of vulnerability to produce a standardized score. Please refer to the CVSS calculator at `https://nvd.nist.gov/vuln-metrics/cvss/v3-calculator`.

The following diagram shows the elements used to calculate it:

CVSS SCORE METRICS

A CVSS score is composed of three sets of metrics (**Base**, Temporal, Environmental), each of which have an underlying scoring component.

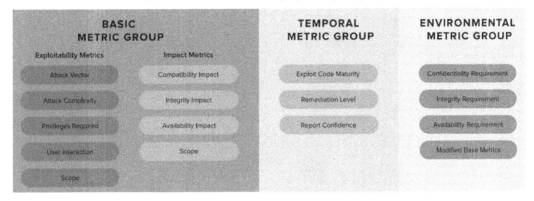

Figure 11.2 – CVSS score matrix
(source: https://www.balbix.com/app/uploads/CVSS-Score-Metrics-Blog-e1592596353700.
png)

As a CISO, the vulnerability is just one aspect of the overall calculation; the threat of that vulnerability impacting the organization (or the **hazard**) or its likelihood of impacting the organization is next. As we translate this to cybersecurity, we need to think about how a vulnerability exists within an environment, for example:

The vulnerablity needs the following aspects to be fully exploitable: it requires physical access to a system, only used on one machine, which is not connected to the larger network, and requires knowledge of the **COBOL** *coding language.*

As we calculate the potential for this example, there is the potential for a major impact due to the results of it being fully exploited.

However, due to the physical access and isolated system, it might be something that occurs rarely and its impact might be limited, netting a low risk rate.

This is not world-ending as there are other aspects of the calculation that we can include:

Risk Rating Matirx					
Impact	Likelihood				
	Rare	**Unlikely**	**Possible**	**Likely**	**Almost certain**
Catastrophic	moderate	moderate	high	critical	critical
Major	low	moderate	moderate	high	critical
Moderate	low	moderate	moderate	moderate	high
Minor	very low	low	moderate	moderate	moderate
Insignificant	very low	very low	low	low	moderate

Figure 11.3 – Risk rating matrix
(source: https://focusergonomics.files.wordpress.com/2013/06/risk_matrix.jpg)

Exposure is the next aspect of the vulnerability that needs to be considered.

For example, do you only have one secret server machine that is not exposed to the internet, physical access, or connected to the network? This might be an offline machine host, with sensitive information such as **crypto wallet keys** or **PKI root CA certificates**. At the other end of the scale, there might be a vulnerability in the base image of your web hosts that are publicly exposed to the internet and scale to millions of machines to handle the increase in web traffic.

The final part of this risk calculation is your **response** to the situation. In terms of cybersecurity, this could be things such as compensating controls (for example, **firewalls**, **network segmentation**, **hardening standards**, and more), **incident response** (**IR**) plans to initiate additional changes, or control of the environment to minimize the effect on the environment. These actions will help reduce, minimize, or eliminate the vulnerability and/or any potential exposure to it.

This is just a simple way to calculate it, but these calculations and any corresponding monetary valuation of the risk will vary based on your knowledge of the environment, the assets, the impact on the business, and exposures.

How would you evaluate or develop a new security program at a new organization?

Your approach could be to use something such as a **strengths, weaknesses, opportunities, and threats (SWOT)** analysis or a threat model either at the organizational level or even at a particular departmental level. I would look at the organization, starting with the business mission, understanding the people and processes currently in place of the business, and how they function to generate and produce results for the organization. Then, I would start to look at internal and external factors that could be a threat to those business people/processes (these could range from inside threats to external threat actors or even regulations and compliance). The following diagram is an example of a SWOT assessment that you might want to consider. In this case, we are considering whether the threat is internal/external to the organization and whether it might have a helpful or harmful impact on the organization:

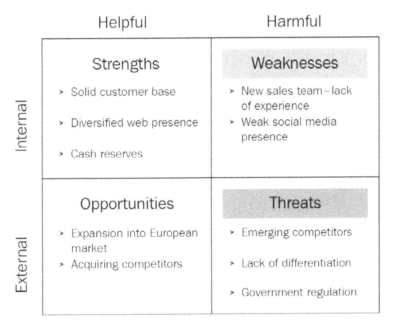

Figure 11.4 – SWOT analysis example

The following diagram is an example of the outcome of a threat modeling exercise, where you would try to understand the potential threats that a specific application or service might face in an organization:

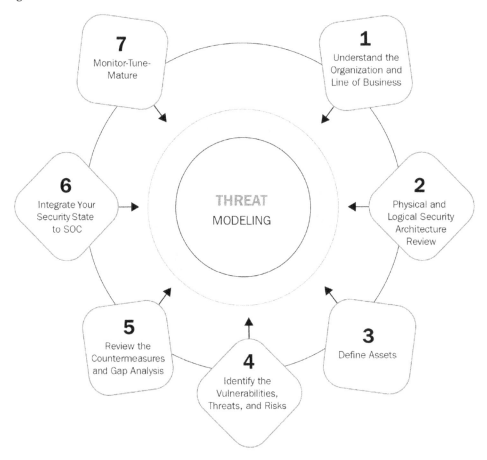

A Successful Threat Modeling Process

Figure 11.5 – Threat modeling framework

Before making any technology recommendations for your organization, consider the existing technologies, current capabilities, and potential capabilities of additional modules, licenses, or features. Additionally, you could do a **return on investment (ROI)** assessment by adding a new solution to the organization or replacing the existing software with a unique solution. Do not forget to think about the people and processes, how these changes will affect them, and whether any user awareness or user training will be needed as a part of the modification process.

A significant part of the CISO's role is becoming a business enabler. It is not the goal of the CISO to implement security for the sake of security; security must focus on the business mission. As you think about implementing security technologies or solutions, you have to think about the people and processes involved, the risk that they help mitigate, the residual risk that is left over, the user or customer friction introduced by the solution, and the value that it adds to the business in the long run. The value or ROI that the solution adds to the company must outweigh the user friction and your ability to mitigate risk to the business. Without this buy-in, end users in the company might not adopt or actively work to circumvent any of the implemented solutions. Alternatively, they might create a business case for their leaders that it is slowing them down too much and costing the business more money than the risk it is mitigating. Situations at this stage end up getting resolved by the business leader with the most political capital and value to the business.

What are your views on a CISO focused on the first line of defense (security operations) versus a CISO focused on the second line of defense – GRC (governance, risk, and compliance)?

Depending on the organization's business mission, the role of the CISO can differ. As a start-up, a company will usually not have a CISO as one of its first couple of employees. Usually, the founders will try to do as much as they can on their own before potentially hiring someone who can help them build the company's structure, and most of the time, they will start with IT first. The IT engineer might implement some security technologies to build the infrastructure. The growth continues; sometimes, their customers, their investors, or their industry start to require more structure around regulations and compliance.

As you can see, based on this growth path, in the beginning, organizations might start with CISOs with a GRC focus and a secondary focus on security operations. As the security program is built and the program matures, the direction of the CISO begins to shift away from the GRC focus toward starting to build out the rest of the security program, including operations (identification and detection), response, and recovery.

Once the program has been built, there is an ingrained conflict of interest regarding what priorities a security program has – operations versus governance. This conflict of interest is very similar to IT versus security, where IT focuses on the availability of the equipment, and security is focused on the confidentiality of the program. In this case, GRC focuses on ensuring that the rules are followed and that there isn't too much risk, while operations try to collect signals and logs on the users, compute resources, and applications to ensure that there isn't any malicious activity.

Sometimes, the amount of data being collected, the speed of that, or how things are being done in order to support the business might result in the collection of too much data in the logs or the deployment of things without following policies and procedures.

Once organizations get to this size or maturity, the CISO might be in charge of operations while reporting to a CIO/CEO or other CxO, and second-line functions such as GRC and auditing might report to another CxO.

For more information on the security line of defense, please visit `https://www.isaca.org/resources/isaca-journal/issues/2018/volume-4/roles-of-three-lines-of-defense-for-information-security-and-governance`.

What are some of the cybersecurity frameworks that you're familiar with? Which one is your favorite and why?

This helps you to show knowledge of the different frameworks within the industry, and whether you, as a candidate, are comfortable with one of the frameworks that the organization is already using. While it is not impossible to introduce a new framework to an organization, the time required to educate all the stakeholders on getting up to speed could potentially be spent elsewhere. One example is the **NIST Cybersecurity Framework** (`https://www.nist.gov/cyberframework`), which allows you to easily communicate what is needed to the leadership and acts as a Rosetta Stone to other commonly used frameworks such as **ISO 27001** (`https://www.iso.org/isoiec-27001-information-security.html`).

What are some of the key components that your third-party risk management program should address?

Some of the key components that should be addressed are the products and systems (both off-the-shelf and open source), the custom applications built by the third party, the cloud services and shared responsibility, and the third-party consultants.

As the SolarWinds supply chain attack, the **Log4j** vulnerability, and breaches in consulting companies, law firms, and governments have demonstrated, third-party risks or supply-chain risks are something that can affect everyone across the board and are something that you need to include as part of your continuous monitoring and risk management. Finding a way to try to effectively mitigate or avoid some of this risk before it happens is a critical part of a **Third-Party Risk Management (TPRM)** process, which can include the reputational status of potential new vendors/suppliers/partners, their external threat posture from externally run scans, and the maturity and effectiveness of their security practices. The last one is tough to do and is usually handled via interviews or questionnaires about their practices before signing the contract or renewal.

As a CISO, what are some ways that you can help map business goals to security decisions for a broad range of stakeholders?

Some of the techniques you can use are tabletop exercises around risk management, informal meetings, and enterprise architecture frameworks such as the **Zachman Framework**, the **Federal Enterprise Architecture Framework (FEAF)**, and **The Open Group Architecture Framework (TOGAF)**.

Tell me about a time when you had to modify an existing security policy and why you had to make the adjustment.

One example could be to determine that the current password policy of the organization did not require long, complex passwords. However, ultimately, this led to privilege escalation and lateral movement in a prior data breach. As the CISO, you then implemented a new policy that required long, complex passwords and that required the use of two-factor authentication, which helped mitigate future attacks.

Tell me about an audit you went through and the outcome.

Years ago, I worked in a consulting (virtual CISO) role at a healthcare organization that underwent an audit that discovered multiple vulnerabilities in the endpoints and some of the web servers. We identified that the critical vulnerabilities in the endpoints were caused by poor patch management and a lack of an antivirus solution. This was coupled with poor password management and open network shares of sensitive data. I worked with their team to implement new policies for passwords, change management, and data security. The network shares were locked down and sensitive data was encrypted. The endpoints were also hardened against attacks. During a follow-up audit 6 weeks later, the auditors identified only a few minor security issues remaining.

Can you tell us about your leadership style?

For this question, the organization is assessing your capabilities to empower your team and delegate appropriately. The worst executive to hire is someone who wants to micromanage everything.

Personally, I combine multiple leadership styles, such as coaching, delegation, visionary, servant, and being emotionally supportive. My approach focuses on setting a high-level objective for the team and then providing them with the space to be creative and play to their own strengths to meet or exceed the objective. I couple this with supporting the team as needed. The result of this approach has been team members who consistently exceed the **key performance indicators (KPIs)** and team members who have invented a new technology that has generated more revenue for the organization.

How do you handle situations where you have to tell someone "no," particularly when they don't report to you?

One of the challenges of being a CISO is that, often, you need to be able to collaborate with stakeholders across the organization, some of whom might not be directly reporting to you. How you deal with these situations is a critical aspect of your ability to collaborate and resolve potential disagreements in differing approaches. Would you work through the *no* with that individual to understand their points of view? Would you escalate the situation to their peer on your level to influence them to have their stakeholder comply? Or, would you potentially come to a compromise in approach with that stakeholder because it was the most efficient way to tackle the situation?

Rather than directly saying *"no"* as it relates to protecting the organization's risk, our compliance posture is saying *"while doing xxx seems like an effective approach, I would ask you to consider xxx."*

The only time I would say *"no"* is when something is deliberately unethical or illegal, and I would remove myself from the situation.

Can you share an experience where you had to collaborate with stakeholders across the organization to implement a security risk management program?

This question assesses your ability to successfully champion security initiatives across the organization.

An example for me is an organization I worked with that had suffered a ransomware attack. I was brought into the organization following the attack and worked with stakeholders across the organization to design, build, and implement a new security risk management program. Key parts of this new program included better detection of and response to incidents, proactive security controls, an organization-wide security awareness training program, scheduled risk assessments, and new security policies. A key aspect of getting buy-in across the organization for this initiative was interviewing each stakeholder to identify what was critical to their team and then identifying the commonalities among teams, along with building out an implementation roadmap that implemented the most critical items first.

How do you train your teams?

Typically, I assess any skill gaps in my team members based on their roles and the needs of the organization. Then, I schedule the more traditional training programs. In some instances, the team will need just-in-time training.

For example, if a major ransomware attack hits the organization and it could have been detected and responded to faster, I will work to train the team in the latest detection and response best practices and work with them to identify any additional security controls that we could implement to protect against similar attacks in the future.

As a CISO, I would also look to develop a pipeline of potential candidates who might be in different parts of the organization. Developing a security champion program helps you drive security awareness and drive the impact of your initiatives deeper into the organization. A by-product of this is that you then have individuals who are knowledgeable of the various lines of business and are also passionate about cybersecurity. Having security-related topics and training as part of the organizational budget versus department budgets also allows you to provide these security champions with training to help prepare them for potential roles in your organization.

Are there any new technologies that you plan to implement after coming on board this organization?

This open question is used to determine your ability to think toward the future. For example, you might answer this question with something such as **Artificial Intelligence** (**AI**) and discuss how you plan to leverage AI capabilities to help improve the organization's security posture. Remember that the effectiveness of an organization's AI program depends on the types of information that you let or do not let it use as part of its analysis. This stems from the organization's data governance posture with data classification and labeling, which allows you to allow or not allow them to be used.

How would you describe a strong information security program?

A strong information security program is one that has a solid foundation of information security policies and procedures. Your information security program needs to align with the business objectives to empower the business and not block it from operations.

How do you measure the effectiveness of your information security program?

You should measure the effectiveness of your information security program by its ability to enable the business to improve productivity and security, without negatively impacting the business operations. For example, enable users to securely share information with clients, as it allows them to become more efficient in their communications to drive sales or customer retention. If the program is being measured in terms of effectiveness, I maintain a scorecard to track it. In a large enterprise, it is important to measure and communicate the effectiveness of programs. Additionally, I use metrics and dashboards that are easy for executives to quickly read.

Let's say you find a development team that has numerous unpatched vulnerabilities that are considered high-risk. How would you influence those individuals to remediate those vulnerabilities?

This question focuses on your ability to influence individuals who are outside your control while also doing it without using authority as a potential driver unless necessary. You should be able to show those outside the organization why something might be important to them and, in turn, impact the organization. Additionally, you should be able to demonstrate to them why resolving it on their own would be more efficient than letting it wait until later. You should also open the door so that, if they need help with resolution, they speak up.

Do you think the size of an organization affects whether security should be outsourced?

Yes; it often costs less to outsource security to a **managed security services provider** (**MSSP**) that already has the in-house talent than to try and hire full-time employees in a smaller organization. The other advantage of outsourcing is the ability to pull in different specialties when you need them. For example, if you need specialists who can conduct advanced malware analysis, it is usually more cost-effective to leverage the existing talent in your third-party service provider.

Larger organizations also outsource, but often, they will have internal security teams.

Tell me about a top executive decision you have made in the past, the circumstances around the decision, and the outcome of that decision.

As an executive, I had to determine the best way to continue scaling the organization while needing to reduce the department's overall spending. I made the difficult decision to remove some of the contract workers we had and temporarily spread their workload around, including assigning some to myself for a short period of time.

How would you describe zero trust to a non-technical stakeholder?

Zero trust is a concept that assumes attackers are already within your network and endpoints. To explain this to a child (or a non-technical stakeholder), I would use the old way of thinking, "Would you trust me with all of your candy because your mom let me in the house?" I would then explain the risk associated with this is that I might take all of your candy and not give you any of it. With zero trust, even though your mom let me inside the house, you would still ask me questions before allowing me access to any candy, and you would have each type of candy segmented from the other types, so I would be unable to get all of your candy in one try.

How does compliance affect your decision-making process?

Compliance will drive the security program for the organization based on the industry. For example, if the organization is in healthcare, the **Health Insurance Portability and Accountability Act (HIPAA)** of 1996, the **Health Information Technology for Economic and Clinical Health (HITECH)** Act, which was enacted as part of the American Recovery and Reinvestment Act of 2009, and the **Health Information Trust (HITRUST)** Alliance are key considerations for compliance, which we will need to address across the organization.

What do you define as the key attributes every CISO should have?

Some of the key attributes I think every CISO should have are effective communication skills, adaptability, effective negotiation skills, active listening skills, and the ability to collaborate effectively with other stakeholders across the organization.

Give me an example of how you balance the cost of your information security program.

Typically, I start with the available budget, review the security controls we need to implement, work with the team to identify the higher-priced options and cost-effective options for controls, and then consider the value of each solution. We conduct a trade-off analysis to determine what we can implement and in what period of time.

Additionally, by tying the security program to business enablement, we can work with other lines of business/departments to enable their initiatives through the security program and have them sponsor a part of the budget for their solutions.

What do you see as some of the emerging risks that enterprises will face?

As of June, 2024, some of the emerging risks that enterprises will face include the increased use of AI in attacks, the increase in ransomware-as-a-service operators, the increase of cybersecurity insurance policy premiums or the lack of insurability, quantum computing and its ability to break current encryption algorithms, and cybersecurity attacks from space assets.

How would you handle a situation in which all your critical systems have been encrypted by a group of hackers? Describe the steps and decisions that need to be made.

Firstly, I want to recommend remaining calm in situations such as this to ensure that you properly consider all the aspects of the situation.

I would recommend letting the system owners verify the health of all the mentioned systems to see how they might be potentially impacted. It should be pretty obvious if the machines or storage systems of these systems are encrypted.

If something seems off, it is time to spin up an investigation team, to start gauging the extent of the event. Then, I would start working with the security team to potentially look at the **indicators of attack (IOAs)**, as well as bringing in threat intelligence to start validating chatter that might be on the internet. As the teams start to potentially discover **indicators of compromise (IOCs)**, you will need to start assessing the severity of the potential incident. (More information about IOAs versus IOCs can be found at `https://www.crowdstrike.com/cybersecurity-101/indicators-of-compromise/ioa-vs-ioc/`.)

Once you have declared the incident, you will want to bring out the **IR** plan to be able to communicate the status to the business and respond to the incident. Hopefully, at this point, there is a well-defined and tested IR plan. The detection and analysis have already started with the IOAs and IOCs. The next phase would, typically, be focused on containment and rededication. You would not want the threat to spread to other systems and cripple the organization further. This containment and rededication should block off communication to these infected systems. **Digital Forensics Incident Responders (DFIRs)** should clone images from these infected systems to start the DFIR investigation process. Part of any IR plan is to ensure that you are able to continually update business leaders, which might include legal and public relations, on the status of the situation.

Backup and restore are likely the next phases of the response. One of the challenges with the backup and restore phases is that many companies only end up doing backups or partial restores. Often, the complexities of backing up and restoring prevent companies from completely validating the backup and restore. Once you have contained the incident within your system, you should start the restore process.

One of the challenges is that threat actors are aware of this process and have now started to target backup processes in order to corrupt or delete backups.

If the backups are unaffected, you should choose the last authentic backup and start the restore process. Once the restore is completed, you should validate that there are no IOCs in the newly restored system. Additionally, the business will also need to validate the restore to ensure that business functionality is fully restored.

Lastly, it is recommended that a post-mortem is conducted so that the teams can learn lessons from the incident. This helps them to update their system configurations regarding detection, IR, backup, restore, and recovery.

How do you communicate the business impact of a breach to other executives? Focus on cost and the cost of solutions to mitigate in the future.

In my experience, cost and compliance are the two main areas of focus. I focus on the cost of the attack or breach, followed by the cost of solutions to mitigate a similar attack in the future, and finally, the compliance concerns the organization has.

Another approach you can take is to align your security program to the business mission and demonstrate how security helps drive business revenue. For example, if a particular attestation or resiliency capability can drive sales or customer retention, work with business leaders to demonstrate that ROI.

How do you empower members of your team in their cybersecurity careers?

I meet with each team member to identify their short- and long-term goals and then identify how I can help each of them within the allotted budget to grow their career.

Using their career drivers, I can work to assign them to projects that align with their career interests and organizational needs. Providing them with ownership of these initiatives allows them to grow in those areas, and they are likely to deliver more value than if they were just assigned to them as part of their role and were of no interest to them.

How do you balance risk to the business and security controls?

I balance the risks to the business and security control implementation by conducting a trade-off analysis of the risk to the daily business operations of each security control and the risk of not implementing those security controls.

How do you stay current on legislation that may affect the organization's security initiatives?

In order to stay current on the legislation that might affect the organization's security strategy, I create Google alerts and monitor **Really Simple Syndication (RSS)** feeds for those topics to ensure that I can see when they are being introduced and the progress they make.

What would you implement to manage endpoints on the network that do not meet minimum security requirements?

An overarching answer here would be zero trust, but specifically, you could use something such as a **network access control (NAC)** policy that blocks devices from the network that do not meet minimum security requirements.

You are the new CISO for a healthcare organization and need to review the policies and procedures of the organization to ensure the organization is compliant with HIPAA and other healthcare laws. What type of security assessment would you perform?

In this situation, you would perform a **regulatory compliance assessment**.

What is your elevator pitch?

Everyone needs to be able to summarize the value that they would bring to an organization in a maximum of one or two sentences. For example, mine is "I translate between people and processes and the technology that helps enable those business missions. I use a consultative and coaching approach to drive overall cybersecurity maturity, the strategy needed, and the people development that achieves organizational success."

As you can see, the role of a CISO can vary dramatically from needing to be tactical regarding the different technical threats that an organization might face to being able to interact and help enable the business mission with counterparts in different departments. The CISO needs to have a varying balance of hard and soft skills to be successful in their role.

Summary

In this chapter, you learned what a CISO is, the average salaries in the US for a CISO, and common questions you might be asked during an interview.

In the next chapter, you will learn about some of the most common behavioral interview questions that are asked in these interviews.

Join us on Discord!

Read this book alongside other users. Ask questions, provide solutions to other readers, and much more.

Scan the QR code or visit the link to join the community.

`https://packt.link/SecNet`

15

Behavioral Interview Questions

This chapter covers some common behavioral interview questions. These questions are grouped by category, and they help employers assess a candidate's soft skills, like **critical thinking** and **teamwork**. You can benefit from learning these questions and practicing their responses, as we authors have experienced many of these questions being asked in cybersecurity interviews verbatim.

In this chapter, the following topics will be covered:

- Why are behavioral questions asked in an interview?
- Common behavioral interview questions

Why are behavioral questions asked in an interview?

Behavioral interview questions are asked to help the interviewer gain insight into how you have handled past situations at work, which can help them see how you might handle similar situations in their company.

Many behavioral questions you are asked in interviews start with phrases similar to the following:

- Describe a situation where you...
- How did you handle [situation]?
- Give me an example of...
- Tell me about a time when you...

Before going into an interview, we suggest you review the job posting and identify the keywords that the employer has listed. This can help you tailor your answers to behavioral interview questions by using some of the keywords from the job description.

Also, I recommend doing a self-assessment to identify 5-10 key skills that you have, and then identify situations where you have used those skills. For example, perhaps you are a good communicator, and you can identify that you used your communication skills to diffuse a difficult situation at a previous job. That is an example you can use to answer one of the behavioral interview questions.

When answering behavioral interview questions, I recommend you use something such as the **P.A.R.** technique, which stands for **Problem**, **Action**, and **Result**.

For the **P.A.R.** technique, you would describe what the problem was, the actions you took, and the result (good or bad). You will also want to mention anyone else that was involved.

Example question:

Tell me about a time when you dealt with a challenging customer.

Example answer using P.A.R.:

In my last job, I worked as a Penetration Tester and dealt with a client who didn't fully understand the value and necessity of penetration testing. They were concerned about the potential disruptions to their organization's workflow and perceived it as an unnecessary expense. During our initial meetings, the client's team had a lot of questions about the process and the methodologies for testing we planned to use.

To address their concerns, I took a proactive approach and did a detailed walkthrough of our testing methodology. I also explained how each step could help identify exploitable vulnerabilities. I also provided examples of case studies from past clients, where our penetration testing uncovered critical vulnerabilities, and how addressing these vulnerabilities with the clients helped improve their overall security posture. The walkthrough and case studies helped to build trust with the new client and show the value of our services. During the penetration test, I ensured open communication with all stakeholders and also ensured everyone was able to understand our final report and recommendations.

The client recognized the value of a penetration test and signed a multiyear contract with our company for our services.

Common behavioral interview questions

Behavioral questions assess a variety of areas to see how you function in a company and contribute to the success of both your team and the organization. I want to stress that your answers to these questions do not have to be targeted toward cybersecurity or IT jobs.

Any situation you have experienced in the past could be used to answer these questions. If you find that you do not have experience that is applicable to a question, inform the interviewer and answer with what you would do if confronted with the scenario presented by the interviewer.

For each of the following questions, an example answer is provided. We did our best to include a variety of potential past roles and examples that are not directly cybersecurity careers; however, we cannot possibly cover every possible reader's personal experience in a single chapter or answer. Please use the example answers as a guide in how you might frame your actual response based on your own personal experience.

Adaptability

The following question is designed to assess your adaptability to changing situations in the workplace:

Describe a time when you had to adapt to changes that were beyond your control. How did you respond to and manage the situation?

During my tenure at a tech startup, a key client unexpectedly pulled out, leading to a sudden loss of revenue. This was a situation completely out of our control. I handled it by quickly assembling a task force to brainstorm and implement strategies to mitigate the impact. We ramped up our efforts to acquire new clients and diversified our service offerings to reduce dependency on any single client. Additionally, I maintained transparent communication with my sales team to keep morale high and ensure everyone was aligned with our new goals. Through these efforts, we were able to stabilize our revenue within a few months.

Ambitiousness

The following questions are intended to assess your level of ambition:

Can you tell me about a project you worked on that was your idea and where you led the implementation?

In my previous role as a cybersecurity consultant, I identified a recurring issue with our clients struggling to maintain secure configurations for their cloud environments. I proposed developing an automated tool that would regularly scan clients' cloud setups for vulnerabilities and compliance issues. I presented this idea to the management, highlighting the potential for increased client satisfaction and retention. Once approved, I led a small team to design and develop the tool. I also coordinated with our sales and marketing teams to market the tool to our clients.

Several clients incorporated the tool and this led to positive feedback from clients and increased revenue.

Describe an important goal you set in the past and how you reached it.

In my previous job, I set a goal to obtain the **Certified Information Systems Security Professional (CISSP)** certification within a year to enhance my credentials and deepen my knowledge of cybersecurity. To achieve this, I created a structured study plan that included daily study sessions, weekend review workshops, and participation in online study groups. I dedicated two hours each evening after work to study the CISSP domains and used practice exams to gauge my progress. Through consistent effort and disciplined study habits, I successfully passed the CISSP exam within the year.

Describe two examples from previous jobs or life experiences that show you are willing to work hard.

First, while working as a junior Penetration Tester, I volunteered to take on additional projects outside my regular duties to gain more hands-on experience. I often worked on these extra projects in my own time, which allowed me to quickly develop my skills and earn a promotion to a senior role within two years.

Second, during my college years, I worked part-time as a security intern while maintaining a full course load. Balancing work and studies required significant effort and time management, but I was committed to gaining practical experience in my field. Despite the demanding schedule, I graduated with honors and received commendations from my internship supervisors for my dedication and performance.

Describe a time when you had to go above and beyond the call of duty in order to get a job done.

In my previous role as an event coordinator, we had a major corporate event where the keynote speaker canceled at the last minute due to an emergency. This could have derailed the entire event, but I took immediate action to find a replacement. I reached out to my network and managed to secure another prominent speaker within a few hours. Additionally, I worked with the other teams to adjust the schedule and make sure the new speaker's requirements were met. I stayed late to ensure everything was set up perfectly. The event went off without a hitch, and the client was extremely pleased with how we handled the unexpected challenge.

Can you tell me about a challenging time at your workplace when a job or task had to be completed? How were you able to focus on completing the job or task?

As a project manager at a marketing firm, we faced a tight deadline to launch a major campaign for a high-profile client. The timeline was reduced due to delays in the approval process.

To ensure we met the deadline, I created a detailed action plan that broke down each task and assigned clear responsibilities.

I set up daily check-ins with the team to monitor progress and quickly address any issues. I also stayed late and worked weekends to provide support and ensure that all deliverables were on track. By maintaining a laser focus and clear communication, we completed the campaign on time, which resulted in a successful launch and a satisfied client.

Can you tell me about a time when you were effective in prioritizing tasks and completing a project on schedule and within budget?

While working as a construction project manager, I was tasked with overseeing the renovation of a commercial building. The project had a strict deadline and budget constraints. I began by breaking down the project into smaller, manageable tasks and prioritized them based on dependencies and deadlines. I implemented project management software to track progress and allocated resources efficiently. Regular progress meetings were held to ensure stakeholders were aligned and any issues were promptly addressed. By maintaining strict control over the schedule and budget, we completed the renovation two weeks ahead of schedule and 5% under budget.

Describe a project you have worked on that demanded a lot of initiative from you.

As a sales manager, I noticed a significant drop in our customer retention rates. Taking the initiative, I proposed and led a customer loyalty program to address the issue. I conducted market research to understand what incentives would be most effective, designed the program structure, and secured buy-in from senior leadership. I then coordinated with the marketing team to launch the program and trained the sales team on how to promote it to customers. The initiative paid off, as we saw a 20% increase in customer retention rates within six months, and it significantly boosted our overall sales revenue.

Describe a time when you worked without close supervision.

I have a few years of experience in graphic design. One time, I was hired by a startup to help develop their brand identity. This client was located in another country, and due to the time difference, I had very little direct supervision from the client on what they wanted for the brand. I took ownership of the project and did research on similar companies to get a better idea of the branding in the marketplace. I kept the client informed about the progress of the project through emails, Slack, and Zoom calls to ensure alignment with the client's overall vision. I delivered a comprehensive brand identity package to the client that included a logo, color palette, and other marketing materials. The client was happy with the final project and left a testimonial that the new branding helped them in a recent product launch.

What impact did you have at your last job?

In my previous role as a customer service manager, I noticed that our response time to customer inquiries was significantly longer than industry standards, which affected customer satisfaction and increased the churn rate. I initiated a project to streamline our customer service processes. This involved implementing a new ticketing system, retraining staff in best practices, and setting up a knowledge base for common issues. Within six months, our response time decreased by 50%, and customer satisfaction scores improved by 20%. This not only enhanced our reputation but also led to increased customer retention and referrals.

What is the most competitive work situation you have experienced? And how did you handle it?

As a sales representative in a highly competitive industry, I was once competing for a major contract against several other top firms. The client was particularly demanding, and every company was putting forth their best offers. To handle this, I focused on building a strong relationship with the client by thoroughly understanding their needs and pain points. I then tailored our proposal to highlight our unique value propositions and how we could address their specific challenges. I also provided case studies and testimonials from similar clients to build credibility. Ultimately, our personalized approach and proven track record won us the contract.

Describe a risky decision you made and the outcome.

While working as a marketing director, I decided to allocate a significant portion of our budget to a new, untested digital marketing campaign targeting a younger demographic. This was a departure from our traditional marketing strategies and carried considerable risk.

I conducted thorough market research and worked closely with a creative agency to develop engaging content tailored to this audience. The campaign included social media promotions, influencer partnerships, and interactive online ads. The decision paid off, as we saw a 35% increase in engagement from the target demographic and a 20% boost in overall sales. The success of this campaign also paved the way for more innovative marketing approaches in the future.

Describe some challenges you faced in your last job and how you handled them.

In my previous role as a project manager for a software development company, one of the biggest challenges we faced was a significant delay in a critical project due to unforeseen technical issues and a key team member leaving the company. To handle this, I immediately re-evaluated the project timeline and identified areas where we could make up lost time without compromising quality. I brought in a temporary contractor with the necessary expertise to fill the gap and reallocated resources to ensure we stayed on track.

I also increased communication with the client to keep them updated on our progress and managed their expectations effectively. By staying proactive and flexible, we were able to complete the project only slightly behind schedule and maintained the client's trust and satisfaction.

Are there projects that you have started on your own recently? And what caused you to start them?

Recently, I initiated a project to develop an internal knowledge-sharing platform at my current organization. I noticed that new employees were spending a lot of time searching for information and learning processes that weren't well documented. This inspired me to create a centralized platform where employees could share best practices, tutorials, and important documents. I collaborated with the IT team to set up the platform and encouraged team leaders to contribute content. The project significantly improved onboarding efficiency and knowledge retention within the company.

Describe some things you have done to invest in yourself and improve your skills.

To continuously improve my skills, I regularly attend industry conferences and webinars to stay updated on the latest trends and technologies. I also complete online courses, which provide me with new tools and skills to improve my cybersecurity work.

Can you tell me about an idea that you came up with during your career and how you applied it?

While working as a retail manager, I noticed that our **inventory management system** was outdated and often led to stock shortages or overstocking. I proposed the idea of implementing a more advanced, automated inventory system that would provide real-time data and analytics. I researched various options, presented the benefits to the management team, and led the implementation process. The new system improved our inventory accuracy, reduced waste, and ensured that we always had the right products in stock, which boosted our sales and customer retention.

How do you handle disagreements with your management team?

I prioritize open and respectful communication. For instance, in my previous role as a marketing coordinator, there was a disagreement about the direction of a major campaign. I took the time to gather data and present a well-researched case for my perspective. I scheduled a meeting with the management team to discuss my findings and listened to their concerns as well. By focusing on facts and maintaining a collaborative attitude, we were able to find a compromise that incorporated elements from both viewpoints, ultimately leading to a successful campaign.

If you have several projects you need to work on, how do you prioritize getting everything done?

When managing multiple projects, I use a combination of prioritization techniques to ensure everything gets done efficiently. First, I assess each project's urgency and importance. Then, I create a detailed schedule with deadlines and milestones for each project. I also communicate regularly with stakeholders to manage expectations and adjust priorities as needed. For example, in my previous role as a marketing manager, I balanced launching a new product campaign while simultaneously preparing for a major industry conference. By breaking down tasks, delegating responsibilities, and maintaining clear communication, I was able to complete both projects on time and within budget.

Analytical thinking

The following questions are designed to test your analytical thinking skills:

Can you provide an example of how you use your analytical abilities?

In my previous role as a sales analyst, I was tasked with identifying trends in our sales data to optimize our sales strategies. I used my analytical abilities to examine large datasets, looking for patterns and anomalies. I developed a comprehensive report that highlighted key sales drivers and areas of improvement. For instance, I discovered that certain products were underperforming in specific regions. By analyzing customer feedback and regional preferences, I recommended targeted marketing campaigns and adjusted inventory levels accordingly. These changes resulted in a 15% increase in sales in those regions within three months.

Developing and using a detailed procedure is important in a job. Tell me about a time when you had to develop and use a procedure to complete a project.

While working as an operations manager at a manufacturing plant, we faced issues with inconsistent product quality. I developed a detailed quality control procedure to address this problem. The procedure included specific steps for inspecting raw materials, monitoring the production process, and conducting final product tests. I trained the team on this new procedure and implemented regular audits to ensure compliance. As a result, we saw a significant reduction in defects and an increase in overall product quality, which improved customer satisfaction and reduced returns by 20%.

Can you walk me through your decision-making process?

My decision-making process involves several key steps:

- Identifying the problem
- Gathering information
- Evaluating alternatives
- Making the decision
- Reviewing the results

For example, as a project coordinator, I had to choose a new software tool for team collaboration. I first identified the specific needs of our team, such as task management and file sharing. I then researched and evaluated various tools, comparing their features, user reviews, and costs. After narrowing down the options, I conducted a trial period with the top candidates and gathered feedback from the team.

Based on this data, I made an informed decision to select the tool that best met our needs and fit our budget. Finally, I monitored the implementation and adjusted it based on user feedback, ensuring the tool was effective and well integrated into our workflow.

Please provide an example of a time when you used good judgment and logic to solve a problem.

In my role as a customer support manager, we encountered a situation where a key client's order was delayed due to a supplier issue. This could have damaged our relationship with the client. Using good judgment and logic, I first assessed the severity of the delay and its potential impact on the client. I then communicated transparently with the client, explaining the situation and apologizing for the inconvenience. To mitigate the issue, I negotiated with an alternative supplier to expedite the delivery and offered the client a discount on their next order as a goodwill gesture. This approach not only resolved the immediate problem but also strengthened our relationship with the client.

Provide an example of when you took a risk to achieve a goal. What was the outcome?

In my previous role as a marketing manager, I took a significant risk by proposing a rebranding campaign for our flagship product, which had been losing market share. This involved changing the product's packaging, logo, and marketing strategy. Despite initial resistance from senior management, I presented a thorough market analysis showing the potential benefits of the rebrand. We conducted focus groups and used their feedback to guide our changes. The rebranding campaign was a success, resulting in a 25% increase in sales over the next six months and revitalizing the product's market presence.

Can you give an example of when using precision was critical for your job?

While working as a quality control inspector at a pharmaceutical company, precision was critical in my job. One instance involved measuring the active ingredient concentration in a batch of medication.

Using highly sensitive equipment, I performed multiple tests to ensure accuracy. Any deviation from the specified concentration could have led to ineffective or harmful medication. My precise measurements and thorough documentation ensured that the batch met all safety and efficacy standards, preventing potential health risks for patients and maintaining the company's reputation for high-quality products.

Tell me about a time when you had to analyze information and make a recommendation.

In my role as a financial analyst, I was tasked with analyzing the potential acquisition of a smaller company. I gathered and reviewed financial statements, market trends, and competitive positioning. After thorough analysis, I created a detailed report highlighting the potential synergies, risks, and financial impact of the acquisition. My recommendation was to proceed with the acquisition but with a focus on specific integration strategies to maximize value. The management team accepted my recommendation, and the acquisition was completed successfully. Over the next year, the company saw a 15% increase in revenue and improved market positioning due to the successful integration.

Can you give me an example of when you had to have attention to detail to complete a task successfully?

While working as a legal assistant, attention to detail was critical when preparing case documents for court. One instance involved compiling evidence and legal briefs for a high-stakes litigation case. I meticulously reviewed each document for accuracy, ensured all citations were correct, and cross-checked the evidence against the claims made. Additionally, I organized the documents in a coherent and logical order for the attorneys. This attention to detail ensured that our legal arguments were presented flawlessly, ultimately contributing to our client winning.

Building relationships

The following questions are designed to see your effectiveness at building relationships:

Give an example of a time when you had to address an angry customer and diffuse the situation.

In my role as a retail store manager, I once dealt with a very upset customer who had received a defective product. The customer was understandably frustrated and demanded a refund.

I calmly listened to their concerns, empathized with their situation, and apologized for the inconvenience. I assured them that we would resolve the issue promptly. I offered them a full refund or an exchange for a new product, along with a discount on their next purchase as a gesture of goodwill. This approach helped to diffuse the situation, and the customer left satisfied with our resolution and continued to shop with us in the future.

Can you tell me about a time when you were not able to build a relationship with someone because they were too difficult to deal with?

In my previous role as a project manager, I worked with a client who was difficult and often unco-operative. Despite my best efforts to understand their needs and maintain open communication, they were consistently critical and unresponsive to our attempts at collaboration. I scheduled regular meetings to try and align our goals, and I provided detailed progress reports to keep them informed. Unfortunately, their negative attitude and unrealistic demands made it impossible to build a productive relationship. Ultimately, we had to escalate the issue to senior management, who reassigned the project to another team with different resources and strategies.

Tell me about a time when you had to quickly establish rapport under difficult conditions.

As a sales representative, I once attended a trade show where I had to engage with potential clients who were skeptical of our product's benefits. The environment was fast-paced, and I had limited time to make an impression. I quickly established a rapport by actively listening to their concerns and asking insightful questions about their specific needs. I then tailored my pitch to address their pain points and demonstrated how our product could provide solutions. I was also honest about how our competitors could solve some of their problems but where the gaps were as well. My ability to connect with them on a personal level and provide relevant information helped build trust and led to sales and long-term client relationships.

What do you think are the keys to establishing successful business relationships? And can you provide an example of how you have used them in your life?

I think the keys to establishing successful business relationships are:

- Communication
- Trust
- Mutual benefit

Open and honest communication ensures that both parties understand each other's needs and expectations. Building trust through reliability and integrity is important for long-term relationships. Finally, ensuring that the relationship is mutually beneficial helps both parties see the value in maintaining it.

In my role as a business development manager, I used these principles to secure a strategic partnership with a major supplier. I initiated regular meetings to discuss our needs and listen to their concerns. By being transparent about our expectations and consistently delivering on our promises, we built a strong foundation of trust.

Additionally, I identified ways our collaboration could benefit both companies, such as co-marketing opportunities and volume discounts. This approach led to a successful partnership that enhanced our supply chain and increased profitability for both parties.

Business systems thinking

The following questions are designed to assess whether you have big-picture thinking and how you navigate office politics:

Describe how your position contributes to the company's overall objectives.

As a marketing manager, my position directly contributes to the company's overall objectives by driving brand awareness and increasing revenue through strategic campaigns. By analyzing market trends and customer data, I develop targeted marketing strategies that align with our business goals. For example, I recently led a digital marketing campaign that focused on increasing our online presence. The campaign resulted in a 30% increase in website traffic and a 20% boost in sales, directly supporting the company's objective of expanding our market share and improving profitability.

Are you a big-picture person or more detail-oriented?

I consider myself a blend of both big-picture and detail-oriented. I excel at understanding the broader goals and vision of a project while also paying close attention to the finer details that ensure its success. For instance, in my previous role as a project manager, I was responsible for overseeing the development of a new software application. I kept the overall objective in mind, which was to enhance user experience and drive customer engagement, while meticulously managing the project timeline, budget, and technical specifications. This balanced approach allowed me to deliver a high-quality product that met both strategic goals and detailed requirements.

Have you experienced a politically complex situation at work? If yes, please describe it and how you came to a positive outcome.

Yes, I encountered a politically complex situation while working as an operations manager. There was a conflict between two department heads over resource allocation, which was having an impact on our project timelines and team morale.

To address this, I facilitated a series of mediation meetings where each party could voice their concerns and needs. I ensured that the discussions remained constructive and focused on finding a solution that would benefit the company. By fostering open communication and encouraging collaboration, we reached a compromise that satisfied both parties. This resolution not only improved inter-departmental relationships but also enhanced overall efficiency and productivity within the company.

Caution

The following questions are designed to assess your ability to use caution when guidelines are not clear and to exercise caution against burnout by performing a self-assessment:

Have you ever worked in a situation where the rules and guidelines were not clear? How did it make you feel?

In my role as a customer service representative, I was once part of a new team assigned to handle a recently launched product. Unfortunately, the guidelines and protocols for managing customer inquiries and issues related to this product were not clearly defined. This ambiguity created confusion and frustration among the team members and made it challenging to provide a consistent service to our customers.

To address this, I took the initiative to collaborate with my colleagues to document our experiences and compile a comprehensive set of guidelines. I organized regular meetings to share insights and best practices, which helped us create a more structured approach to handling customer queries. While the initial lack of clarity was stressful, taking proactive steps to develop clear guidelines ultimately improved our team's efficiency and customer satisfaction.

Can you tell me about a time when you demonstrated too much initiative on a project?

During my tenure as a marketing coordinator, I was excited about a new social media campaign we were planning. I took the initiative to create a detailed content calendar, design graphics, and draft posts without first consulting the rest of the team. I was eager to get things moving quickly, but my hasty approach meant that I overlooked some critical steps, such as aligning the campaign with our overall marketing strategy and getting necessary approvals from the management team.

As a result, some of the content I created had to be revised, and the campaign launch was delayed. This experience taught me the importance of balancing initiative with collaboration and proper planning.

I learned to ensure that all stakeholders are involved and that projects align with the broader objectives of the team and organization. Moving forward, I made it a point to involve key team members from the beginning and seek regular feedback to ensure that we were all on the same page.

Communication

The following questions are designed to assess your communication skills:

Can you tell me about a situation where you were able to effectively "read" another person and tailor your actions to your understanding of that person's needs?

While working as a customer service representative, I encountered a customer who seemed very frustrated and upset about a billing issue. I could tell from her tone and body language that she was feeling overwhelmed. Instead of following the usual script, I decided to take a more empathetic approach. I calmly acknowledged her frustration and assured her that I was there to help. By actively listening and validating her feelings, I was able to understand the root of her concern. I then explained the billing process clearly and concisely, providing her with a detailed breakdown of the charges. By tailoring my approach to her emotional state, I was able to defuse the situation and resolve her issue, leaving her satisfied and appreciative of the personalized assistance.

Can you share an experience where you didn't feel like you communicated effectively? What happened and how did you resolve the situation?

Early in my career as a project coordinator, I was tasked with leading a team meeting to discuss upcoming deadlines and deliverables. During the meeting, I provided a lot of detailed information but didn't notice that several team members looked confused. I assumed everyone understood and didn't ask for feedback or questions. Later, I realized that some team members were unclear about their tasks, leading to delays in the project. To resolve this, I scheduled a follow-up meeting where I encouraged open communication and asked each team member to repeat their understanding of their tasks. I also provided written summaries of the key points and deadlines discussed. By fostering a more interactive and supportive communication environment, I ensured everyone was on the same page, which helped get the project back on track.

How have you used effective communication to strengthen a relationship?

As a sales executive, I worked with a long-term client who had recently expressed dissatisfaction with our services.

To address this, I scheduled a face-to-face meeting to discuss their concerns. I made sure to listen attentively, asking clarifying questions to fully understand their issues. By acknowledging their concerns and showing genuine interest in finding a solution, I was able to rebuild trust.

How would you communicate a difficult or unpleasant idea to your manager?

If I needed to communicate a difficult or unpleasant idea to my manager, I would first gather all relevant facts and data to support my position.

For example, if our project were falling behind schedule, I would prepare a detailed report outlining the specific reasons for the delay, potential impacts, and possible solutions. During the conversation, I would be honest and straightforward and would present the issue clearly. I would explain the steps already taken to address it and propose a plan for moving forward. By focusing on solutions and maintaining a positive, proactive attitude, I would aim to facilitate a constructive discussion and demonstrate my commitment to resolving the issue.

What is the most important presentation you have completed?

The most important presentation I completed was during my role as a business development manager, where I pitched a strategic partnership to a potential client that could have an impact on our company's growth. The presentation involved a comprehensive analysis of market trends, detailed projections of mutual benefits, and a well-structured plan for collaboration.

During the presentation, I focused on clearly communicating the value proposition and addressing potential concerns. My effort paid off, as the client was impressed and agreed to the partnership, which resulted in a substantial increase in our market reach and revenue.

Can you give an example of how you successfully communicated with someone who didn't like you or who you didn't like?

In my previous role as a team leader, I had a colleague who I knew wasn't particularly fond of me due to a past disagreement. We had to collaborate on a project, so I focused on professional and transparent communication. I made a point to listen actively to their ideas and show respect for their expertise. During our meetings, I kept the discussion centered on the project's objectives and how we could achieve them together. By maintaining a positive and constructive attitude, I was able to foster a more cooperative working relationship. Our communication improved and we successfully completed the project.

Describe a time you had to sell an idea to your co-workers or a manager. What was your process? And were you successful in selling the idea?

In my role as a marketing coordinator, I once proposed a new social media strategy to increase our brand's online presence. To sell this idea, I first conducted thorough research and collected data showing the potential benefits.

I prepared a detailed presentation that included market analysis, case studies, and projected ROI. I also anticipated potential objections and prepared responses to address them. During the presentation to my co-workers and manager, I emphasized the strategic alignment with our overall marketing goals and demonstrated how the new approach could drive engagement and sales. My preparation and clear communication helped gain their buy-in, and the strategy was implemented, leading to a 30% increase in our social media engagement within three months.

Have you ever had a subordinate whose performance was consistently below average? What did you do to improve their performance?

Yes, I once had a team member whose sales performance was consistently below average. I started by having a one-on-one meeting with them to understand any underlying issues and provide feedback. We identified that a lack of product knowledge and confidence in closing deals were key areas for improvement. I created a personalized development plan that included additional training sessions, role-playing exercises, and regular check-ins to monitor progress. I also paired the individual with a mentor who was a top performer on the team. Over time, I noticed a significant improvement in their performance, and they began to meet and even exceed their sales targets.

How do you communicate critical information to your team?

When communicating critical information to my team, I prioritize clarity and timeliness. I use multiple channels, such as email for detailed information, and follow up with a brief team meeting to ensure everyone understands the message and its importance. For example, during a product launch, I sent out an email outlining the timeline, key responsibilities, and potential challenges. I then held a team meeting to discuss the plan, answer questions, and address any concerns. I also encouraged team members to reach out if they needed further clarification. This approach ensured that everyone was on the same page and could act on the information effectively.

How do you keep your manager informed about your work?

I keep my manager informed about my work through regular updates and transparent communication. I provide weekly status reports that outline my progress, any challenges encountered, and the next steps.

Additionally, I schedule bi-weekly one-on-one meetings to discuss ongoing projects in more detail and address any urgent issues. For instance, while working on a critical project, I maintained a shared project dashboard that my manager could access at any time to view real-time updates. This consistent communication ensures that my manager is always aware of my work and any potential roadblocks, allowing for timely interventions if needed.

What types of communication situations cause you difficulty? Please provide an example.

One type of communication situation that can cause me difficulty is addressing underperformance issues with team members. I once had to discuss consistent tardiness and missed deadlines with a team member who was otherwise well liked and hardworking. To address the situation, I scheduled a private meeting where I carefully outlined the specific issues and their impact on the team's performance.

I focused on being empathetic and supportive, aiming to understand any underlying issues they might be facing.

While it was uncomfortable, approaching the conversation with empathy and a solution-oriented mindset helped us come up with an action plan to improve their performance.

Tell me about a time you gave a successful speech. How did you prepare for it?

During my last role as a sales manager, I was invited to speak at a regional sales conference about innovative sales strategies. To prepare, I first researched and identified key trends and data that would be relevant and valuable to the audience.

I then structured my speech to include an engaging introduction, a detailed exploration of successful strategies, and practical tips they could implement immediately. I practiced multiple times in front of a mirror and with a few colleagues to get feedback. On the day of the conference, I delivered the speech confidently and received positive feedback from attendees who appreciated the actionable insights and clear delivery.

Describe a time you used your communication skills effectively to make a point.

In my role as a team leader, I noticed that our meetings were often unproductive due to a lack of clear agendas and objectives. I decided to address this by clearly communicating the importance of structured meetings to my team. During a team meeting, I presented data showing the time wasted in our previous meetings and outlined a new approach with specific agendas and time allocations for each topic. I used clear examples of how this structure could improve efficiency and productivity. By effectively communicating the benefits and providing a clear plan, I was able to convince the team to adopt the new meeting structure, which resulted in more focused and productive meetings going forward.

Tell me about a time you were effective in giving a workshop.

While working as a sales manager, I was asked to lead a workshop on effective negotiation techniques for our sales team.

To ensure the workshop was impactful, I designed interactive activities and real-life role-playing scenarios that allowed participants to practice the skills in a safe environment with immediate feedback. I prepared detailed handouts that summarized key points and provided actionable tips. During the workshop, I encouraged active participation and open discussion, which helped keep everyone engaged. The feedback was overwhelmingly positive, with many team members reporting improved confidence in their negotiation abilities, which subsequently led to an increase in successful deals for our organization.

Give an example of a time when you needed to speak up during a project and the outcome.

During a major product launch, I noticed that our marketing materials were not highlighting a key benefit that could attract a new customer segment. Despite being relatively new to the team, I felt it was important to bring this up.

I gathered data to support my observation and presented it in our next meeting, explaining the potential benefits of emphasizing this benefit. My suggestion was initially met with some resistance, but after discussing the data and potential impact, the team agreed to incorporate the change.

This adjustment led to a 15% increase in interest from the targeted customer segment, contributing to the product's successful launch.

Can you tell me about a time when you had to be assertive? What was the outcome?

As a project coordinator, I was once tasked with ensuring that all team members met their deadlines for a critical project. One team member repeatedly missed their milestones, which jeopardized the project's timeline. I had to be assertive and address the issue directly. I scheduled a private meeting with the individual to discuss the importance of their contributions and the impact of their delays. I provided clear expectations and set a firm deadline for their next deliverable. By being assertive yet supportive, the team member understood the urgency and made improvements in their performance, helping us complete the project on time.

Describe a time you used effective written communication to make a point.

While working as a customer service supervisor, I noticed a recurring issue with how some team members handled customer complaints. To address this, I wrote a detailed email to the team outlining the problem, its impact on customer satisfaction, and proposed solutions. I used clear, concise language and included examples to illustrate my points. I also attached a revised complaint-handling procedure for their reference.

The email prompted a productive discussion during our next team meeting, and the new procedures were quickly adopted. As a result, we saw a significant improvement in our customer satisfaction scores.

Tell me about a challenge you faced while coordinating a project with other teams.

While managing a cross-functional project involving both the marketing and product development teams, I encountered a significant challenge in aligning their priorities and timelines. The marketing team needed early access to product details for their campaign, while the development team was still finalizing the product features.

To address this, I organized a joint meeting where both teams could voice their concerns and needs. I facilitated a discussion to find common ground and proposed a phased approach, where key features would be shared as they were finalized. This compromise allowed the marketing team to begin their work without delays, while the development team could continue to refine the product. This coordination ensured the project stayed on track and met its deadlines, ultimately resulting in a successful product launch.

What have you done in your career to improve your verbal communication skills?

With this question, the interviewer is trying to determine whether you are willing to continue your education and go outside of your comfort zone to advance your career.

To improve my verbal communication skills, I have actively sought out opportunities to speak publicly and engage in discussions. For instance, I joined a local Toastmasters club, where I regularly participated in speaking exercises and received constructive feedback from peers. Additionally, I volunteered to lead team meetings and present at company events, which helped me gain confidence and improve my speaking abilities. I also took a course on effective communication at my local college, which provided valuable techniques for structuring my messages and engaging my audience.

How have you persuaded people with your documentation?

In my previous role as a project manager, I often had to create detailed project proposals to secure buy-in from stakeholders. One instance involved persuading the executive team to invest in a new software tool. I crafted a report that included a thorough analysis of the current challenges, the benefits of the proposed tool, cost analysis, and a clear implementation plan.

I used data and case studies to support my position and included testimonials from other companies that had successfully adopted the tool. The well-structured and evidence-backed document convinced the executives of the tool's value, and we received the approval to proceed with the purchase.

What are the most challenging documents you have created?

One of the most challenging documents I worked on was a comprehensive strategic plan for a major product launch. The document required input from multiple departments, including marketing, sales, product development, and customer support. Coordinating and integrating diverse perspectives and data into a coherent and compelling strategy was a significant challenge. I had to ensure that all the stakeholders' concerns were addressed and that the plan was both actionable and aligned with the company's overall objectives. The meticulous planning, multiple revisions, and constant communication with various teams made it a demanding task, but the final document was instrumental in the successful launch of the product.

What is your process for preparing written communication?

My process for preparing written communication involves several key steps to ensure clarity and effectiveness. These steps are:

- *Defining* the purpose and audience of the communication, which helps tailor the content and tone appropriately.

- *Gathering* all relevant information, organizing it logically, and creating an outline to structure the document.

- *Creating a draft* of the content, focusing on clear and concise language, and using headings and bullet points to enhance readability.

- *Reviewing and editing* the document for accuracy, coherence, and grammar. I also seek feedback from colleagues to ensure the message is clear and effective.

- *Proofreading* the document one last time before sending it out to ensure it is polished and professional.

Conflict resolution

The following question assesses your ability to resolve conflicts in a way that benefits everyone involved:

Describe a time you had a conflict with your supervisor and how you handled it.

I once disagreed with my supervisor about the marketing campaign direction. They wanted to stick with traditional methods, while I believed a more digital-focused approach would yield better results. To handle this conflict, I gathered data and case studies to support my viewpoint. I scheduled a meeting with my supervisor and presented my findings calmly and respectfully. I made sure to listen to their concerns and acknowledged the value of their experience.

By focusing on the potential benefits and showing respect for their perspective, we were able to reach a compromise that incorporated both traditional and digital strategies, ultimately leading to a successful campaign that increased sales by 11%.

Describe a time when you took accountability during a conflict with another person and how you resolved the conflict to benefit both parties.

In my previous role as a sales team leader, I had a conflict with a colleague in the marketing department regarding the strategy for a new campaign.

The marketing team had created promotional materials that I felt did not accurately represent the product's key features, which I believed would affect our sales performance.

The disagreement escalated during a meeting, and it became clear that our teams were not on the same page. Realizing that the conflict was impacting our progress, I decided to take accountability for my part in the escalation. I approached my colleague after the meeting and acknowledged that I could have communicated my concerns more constructively. I apologized for any frustration I had caused and expressed my willingness to collaborate more effectively. To resolve the conflict, I suggested we schedule a follow-up meeting with a clear agenda and include both our teams to foster a collaborative environment.

During the meeting, we encouraged open dialogue and actively listened to each other's perspectives. We identified common goals and worked together to revise the promotional materials in a way that highlighted the product features more accurately while still aligning with the marketing strategy. The revised campaign was well received, and our collaboration improved, leading to a successful product launch and better interdepartmental relations.

Customer orientation

The following questions assess your ability to interact with clients:

Describe how you handle problems with customers.

Handling problems with customers requires a calm and empathetic approach. In my previous role as a customer service manager at a retail company, I dealt with a situation where a customer received a damaged product.

The customer was understandably upset. I first listened carefully to their complaint without interrupting to understand the full extent of the issue. I then apologized sincerely for the inconvenience and reassured them that I would resolve the problem promptly. I offered a replacement product to be shipped immediately at no additional cost, along with a discount on their next purchase as a gesture of goodwill. By actively listening and providing a quick, fair solution, I was able to turn a negative experience into a positive one, resulting in a satisfied and loyal customer.

How do you go about establishing rapport with a customer? What do you do to gain their confidence? Give an example.

Establishing rapport with a customer involves showing genuine interest in their needs and building trust through reliable service. When I worked as a sales representative for a tech company, I made it a point to research my customers' businesses before meetings.

During one instance, I met with a potential client who was hesitant about investing in our software solution. I asked specific questions about their business challenges and listened attentively to their concerns. I shared success stories of similar clients and how our software had benefited them.

I also provided a customized demonstration highlighting features that addressed their specific needs. By showing that I understood their business and providing tailored solutions, I gained their confidence, and they decided to proceed with a trial of our software, which later led to a long-term partnership.

What is your process of determining how to handle a difficult customer?

When dealing with a difficult customer, my process involves several steps. First, I listen carefully to their concerns without interrupting, showing empathy and understanding their perspective. I then clarify and summarize their issue to ensure I have understood it correctly.

Next, I evaluate the situation, considering company policies and the customer's history with us. I propose a fair and reasonable solution that aligns with our policies and addresses the customer's concerns. For example, when a customer was unhappy with a delayed delivery, I listened to their frustration, acknowledged the inconvenience, and offered a solution by expediting a replacement order and providing a discount on their next purchase. This approach helped resolve the issue and retained the customer's business.

What have you done to improve relations with your customers?

Improving customer relations often involves proactive communication and exceeding their expectations. In my role as an account manager at a logistics firm, I noticed that some clients felt out of the loop regarding their shipment statuses.

To address this, I implemented a regular update system where clients received timely email notifications about their shipments' progress. Additionally, I scheduled quarterly review meetings to discuss their experiences and gather feedback. During one of these meetings, a client mentioned their growing need for expedited shipping options. I worked with our operations team to develop and offer a new premium service that catered to their requirements. These efforts improved not only our communication and service but also the stickiness of clients.

Decision-making

The following questions are designed to assess your effective decision-making abilities:

Discuss an important decision you have made regarding a task or project at work. What factors influenced your decision?

In my role as a project manager at a marketing agency, I was tasked with deciding whether to extend a campaign timeline due to unforeseen delays or push the team to meet the original deadline.

The factors influencing my decision included the potential impact on client satisfaction, the quality of the deliverables, and the team's workload. After analyzing the situation, I consulted with the team to understand their capacity and challenges. I also reviewed the client's goals and expectations. Ultimately, I decided to extend the timeline slightly, ensuring high-quality deliverables and avoiding team burnout. I communicated this decision transparently to the client, who appreciated the focus on quality and was satisfied with the revised timeline.

Have you ever made a poor decision at work? How did you recover?

Early in my career as a sales associate, I made the poor decision to promise a customer a delivery date without first checking our inventory levels. When I realized we couldn't meet the promised date, I immediately contacted the customer to apologize and explain the situation.

I offered a solution by expediting the order at no extra cost and provided a discount on their next purchase as compensation. Additionally, I worked with our inventory team to implement a more reliable stock-checking system before making delivery commitments. This experience taught me the importance of verifying information before making promises and improved our process to prevent similar issues in the future.

Describe a time when you had to make a quick decision.

As a shift supervisor at a busy restaurant, there was an instance when a critical kitchen appliance broke down during peak hours. I had to make a quick decision to ensure service continuity.

I immediately reassigned staff to cover the affected tasks manually and called a repair technician. To minimize disruption, I adjusted the menu to focus on dishes that didn't require the machine and communicated transparently with customers about the temporary change. This quick decision helped maintain service efficiency and customer satisfaction despite the unforeseen issue.

Can you provide an example of a time when you didn't have enough information to make a decision? What did you do?

As a product development coordinator, I once faced a decision on whether to proceed with a new product feature based on limited market data.

Instead of making an uninformed decision, I requested a short extension to gather more information. I organized a series of customer surveys and competitor analyses to better understand the market demand and potential impact of the feature. The additional data provided clarity and allowed us to make an informed decision to proceed with the feature, which ultimately contributed to the product's success.

What is your process for making important decisions?

My process for making important decisions involves several key steps to ensure thoroughness and effectiveness. First, I gather all relevant information and data to understand the full scope of the situation. Next, I identify the key objectives and criteria that the decision needs to meet. I then brainstorm and evaluate potential options, considering the pros and cons of each. I also consult with colleagues or experts to gain different perspectives and insights. After weighing all the factors, I make a decision that aligns with the overall goals and values of the organization. Finally, I monitor the outcomes and remain flexible to adjust if necessary.

How do you involve your manager or others when you make a decision?

Involving my manager and other stakeholders in the decision-making process is important for ensuring alignment and gaining buy-in. I start by clearly defining the decision to be made and identifying who needs to be involved.

I then organize meetings or discussions to present the gathered information, potential options, and my initial recommendations. I encourage open dialogue and actively seek input and feedback from my manager and other stakeholders. This collaborative approach not only helps refine the decision but also builds consensus and buy-in. For example, when planning a new marketing campaign, I involved the sales, product, and customer service teams to ensure the campaign was comprehensive and well supported by other teams.

What is your process for identifying whether a decision you make is beneficial?

To identify whether a decision is beneficial, I establish clear criteria and **key performance indicators (KPIs)** that align with the intended outcomes of the decision. I regularly monitor and measure these KPIs to track progress and assess the impact of the decision. I also gather feedback from relevant stakeholders to understand their perspectives on the decision's effectiveness.

Additionally, I conduct periodic reviews to compare the actual outcomes with the expected results and adjust if needed.

Tell me about a time when you defended your decision successfully even though key stakeholders were initially opposed to it.

In my role as a project manager, I once proposed implementing new project management software to improve team collaboration and efficiency. Some key stakeholders were initially opposed to the idea, citing concerns about the cost and the learning curve associated with the new tool. To defend my decision, I conducted a thorough cost-benefit analysis that highlighted the long-term savings and productivity gains from using the software.

I also arranged a demonstration session where the team could see the software's features and benefits firsthand. Additionally, I proposed a phased implementation plan with training sessions to address the learning curve concerns. By presenting concrete evidence and a clear implementation strategy, I was able to gain the stakeholders' support and successfully implement the new software, which ultimately enhanced our project management capabilities and overall productivity.

Describe a problem you have had when coordinating technical projects.

While coordinating a website redesign project at my previous job, we faced a significant problem when the development team encountered unforeseen compatibility issues with our existing database. This problem threatened to delay the project timeline and increase costs. To address this, I organized a meeting with the development team and our database administrators to identify the root cause of the issues. We brainstormed potential solutions and decided to implement a temporary workaround to maintain project momentum while working on a more permanent fix.

I also adjusted the project timeline and communicated the changes transparently with all stakeholders, ensuring that everyone was aware of the new plan and expectations. This collaborative approach helped us overcome the technical challenges and complete the project successfully.

What was your most difficult decision in the last six months and why was it difficult?

In the last six months, the most difficult decision I faced was whether to downsize our team in response to budget cuts. This decision was challenging because it involved not only financial considerations but also the impact on team morale and individual livelihoods. I conducted a thorough analysis of our budget and performance metrics, consulted with senior management, and considered alternative cost-saving measures. Ultimately, I decided that a small reduction in team size was necessary to ensure the company's long-term sustainability.

To handle the situation with empathy and transparency, I communicated the reasons for the decision to the affected employees and provided support in the form of severance packages and assistance with job placement. This approach helped to mitigate the negative impact of the layoffs.

Can you tell me your process for making highly technical decisions?

When making highly technical decisions, my process involves several key steps to ensure accuracy and effectiveness. First, I gather all relevant technical data and information related to the decision. I consult with subject matter experts to gain insights and validate the information. Next, I analyze the potential options and their implications, considering both short-term and long-term effects. I often use decision-making frameworks or tools, such as **SWOT analysis** or **cost-benefit analysis**, to evaluate the options objectively.

Once I have a clear understanding of the best course of action, I communicate the decision and its rationale to all stakeholders, ensuring they are informed and aligned. For example, when deciding to upgrade our company's IT infrastructure, I followed this process to choose the most suitable and cost-effective solution, which ultimately improved our operational efficiency and supported our growth objectives.

Delegation

The following questions assess your ability to delegate tasks to others:

Do you consider yourself a macro or micromanager?

I consider myself a macromanager. I believe in providing my team with the autonomy they need to perform their tasks while offering guidance and support when necessary. For example, as a marketing manager, I set clear objectives and expectations for a campaign but allowed my team the creative freedom to develop the strategies and materials. This approach not only empowers team members to take ownership of their work but also builds innovation and collaboration.

By focusing on the big picture and trusting my team to handle the details, we can achieve our goals effectively.

How do you make decisions when delegating work to your team?

When deciding to delegate work, I consider several factors. First, I assess the strengths and expertise of each team member to ensure the task aligns with their skills and development goals. Next, I evaluate the workload and capacity of each team member to avoid overburdening anyone.

I also consider the complexity and importance of the task, ensuring that critical tasks are delegated to experienced team members. For example, during a product launch, I delegated specific marketing tasks to team members based on their expertise in social media, content creation, and market analysis. This allowed each person to contribute their best work while maintaining a balanced workload across the team.

What is the biggest mistake you have made when delegating tasks?

The biggest mistake I made when delegating tasks was assuming a team member had the necessary skills for a task without providing adequate support or training. Early in my career, I assigned a complex data analysis project to a junior team member without realizing they were not fully proficient in the required software. The project was delayed, and the results were not up to standard. I took accountability for the oversight, provided the necessary training, and closely monitored their progress on similar future tasks. This experience taught me the importance of ensuring team members have the right skills and support before delegating complex tasks.

Can you tell me about your biggest success when delegating tasks?

My biggest success in delegating tasks occurred during a major event-planning project. I delegated specific responsibilities such as vendor coordination, marketing, and logistics to team members based on their strengths and interests. I provided clear guidelines and expectations but trusted them to handle the details. This delegation allowed me to focus on overseeing the overall project and addressing any high-level issues that arose. The event was a tremendous success, with positive feedback from attendees and stakeholders. Each team member felt valued for their contributions, which boosted team morale and demonstrated the power of effective delegation.

Detail orientation

The following questions are designed to assess your attention to detail:

Describe a situation where you had the option to leave the details of a project to someone else or handle them yourself and what your decision was and why.

While managing a large marketing campaign, I had the option to delegate the detailed work of content creation and scheduling to a team member or handle it myself. Given the team's workload and the tight deadline, I decided to delegate these tasks to a team member who had shown exceptional skill in content creation. I made this decision to focus on overseeing the campaign strategy and ensuring all parts were cohesive and aligned with our objectives. By trusting my team member with the details, I was able to maintain a high-level view of the project, address any overarching issues, and ensure the campaign's success.

Reflecting on previous jobs, overall, have those jobs required low attention to detail, a moderate amount of attention, or a high amount of attention to detail to be successful in the job?

Reflecting on my previous jobs, they have generally required a high amount of attention to detail to be successful. For example, in my role as a project manager for a software development company, precision was important for ensuring that project specifications were met, deadlines were adhered to, and client expectations were exceeded. Attention to detail was also essential in creating accurate project documentation, conducting thorough quality checks, and maintaining clear communication with stakeholders. This focus on detail has consistently helped in delivering high-quality outcomes and building trust with clients and other stakeholders.

Can you share a situation where you found it challenging to handle the details of a project?

One challenging situation involved coordinating a company-wide system upgrade. The project required attention to detail, from scheduling downtime to ensuring data integrity and training employees on the new system.

The sheer volume of details was overwhelming initially, and I found it challenging to keep track of all the moving parts. To manage this, I implemented a detailed project plan using a task management tool, which allowed me to break down the project into manageable tasks and assign responsibilities to team members. Regular check-ins and status updates helped ensure nothing was overlooked. Despite the complexity, the project was completed successfully, with minimal disruption to our operations.

Can you describe a situation where you found attention to detail was not important?

There was a situation during a brainstorming session for a new marketing campaign where attention to detail was not the primary focus. The session's goal was to generate a wide range of creative ideas without immediately worrying about feasibility or specific details. By encouraging the team to think broadly and creatively without getting bogged down in the minutiae, we were able to come up with innovative concepts that we later refined and developed.

Employee development

This question is designed to assess your ability to build out employee development training:

Have you ever developed or enhanced a training program? If yes, please tell me your process for building one.

Yes, I developed a training program for new sales associates at my previous job. The process began with a thorough needs assessment to understand the skills and knowledge gaps of new hires. I interviewed current employees and managers to gather insights into the most critical areas needing improvement. Based on this feedback, I designed a comprehensive curriculum that included both theoretical and practical components. I created detailed training materials, including manuals, presentations, and interactive exercises. I also incorporated assessments and feedback loops to ensure continuous improvement. The program was piloted with a small group, and their feedback was used to make further refinements. Ultimately, the enhanced training program significantly improved the onboarding experience and reduced the time it took for new hires to become productive.

Evaluating alternatives

The following questions are designed to assess your process for evaluating alternate options:

Describe a situation where you had multiple options to choose from. How did you reach a decision about which options to choose?

In my role as a project manager, I once had to choose between three vendors for a major software implementation. Each vendor had its strengths, and the decision was critical for the project's success. I created a detailed comparison matrix evaluating each vendor on key criteria such as cost, experience, technical capabilities, and support services. I also sought input from my team and other stakeholders to get their perspectives.

After analyzing all the data and considering the long-term impact, I chose the vendor that offered the best balance of quality and cost-effectiveness. The decision was well received, and the implementation went smoothly, meeting all our objectives.

Were there any alternate options you developed as part of your review process?

Yes, during the vendor selection process for the software implementation project, I developed alternate options by considering hybrid solutions that combined services from different vendors.

I explored the possibility of using one vendor for the core software and another for support services to leverage the strengths of each. This approach allowed us to negotiate better terms and ensured we had backup plans if the primary option fell through.

Ultimately, while we proceeded with a single vendor for simplicity and integration benefits, having these alternate options provided us with a safety net and increased our negotiating power.

Can you share a major decision you have made in the past 12 months and how you evaluated the other options that were available?

A major decision I made in the past 12 months was to expand our product line to include a new range of eco-friendly packaging. The decision involved evaluating several options: developing the products in-house, partnering with an external manufacturer, or acquiring an existing company with expertise in this area. I conducted a thorough market analysis to understand demand and potential profitability. I also evaluated the cost, time, and resources required for each option. After presenting a detailed report to the executive team and discussing the potential risks and benefits, we decided to partner with an external manufacturer. This allowed us to quickly bring the product to market while minimizing upfront investment and leveraging the partner's expertise.

Are there types of decisions that you find more difficult to make than others?

I find decisions involving significant personnel changes, such as hiring or letting someone go, more difficult than others. These decisions impact not only the individuals involved but also the team's dynamics and morale. For example, during a restructuring process, I had to decide which positions to eliminate to streamline operations. To make this decision, I conducted a thorough performance and contribution analysis for each role, considered the long-term strategic goals, and sought input from senior management. Despite the difficulty, I approached it with empathy and transparency, ensuring affected employees received adequate support and clear communication throughout the process.

Flexibility

The following questions are designed to assess your flexibility on the job:

Describe how you adjust your management approach if you see it's not working to improve performance.

When I see that my initial management approach isn't yielding the desired results, I first seek feedback from the team to understand their perspective. For example, while managing a project team, I noticed that the initial top-down approach wasn't fostering collaboration.

After gathering feedback, I shifted to a more participatory style, involving team members in decision-making processes and encouraging open communication. I also implemented regular team meetings to discuss progress and challenges openly. These adjustments created a more inclusive environment, improved team morale, and significantly enhanced overall performance.

What is your process for working through obstacles that you face in projects?

My process for working through obstacles in projects involves several key steps. First, I identify and clearly define the obstacle. Then, I analyze the root cause and gather relevant information from all stakeholders involved. For example, during a project to launch a new product, we encountered a major supplier issue. I convened a meeting with the team to brainstorm potential solutions, weighing the pros and cons of each. We decided to temporarily switch suppliers and renegotiate terms to ensure timely delivery. I also developed a contingency plan to address any future disruptions. This systematic approach allowed us to overcome the obstacle and successfully launch the product on time.

Follow-up and control

The following questions are designed to assess your process for tracking metrics:

What is your process for tracking projects that you have delegated to others?

My process for tracking delegated projects involves several steps. First, I set clear expectations and deadlines for each task. I use project management software like **Monday**, **Trello**, or **Asana** to assign tasks and monitor progress. Regular check-ins, either through weekly meetings or status updates, help ensure that everyone stays on track. For example, in my last role, I managed a marketing campaign where I delegated tasks to different team members. By using Asana, I could easily track each person's progress, provide timely feedback, and address any roadblocks they encountered. This systematic approach ensured that the project was completed on time and met all our objectives.

What is your process for collecting data for performance reviews?

My process for collecting data for performance reviews includes a combination of quantitative and qualitative methods. I start by gathering metrics related to the employee's **key performance indicators (KPIs)**, such as sales numbers, project completion rates, or customer feedback scores. I also collect feedback from peers, subordinates, and supervisors through anonymous surveys or direct interviews. Additionally, I review the employee's previous performance reviews and self-assessments.

For instance, during annual reviews at my last job, I compiled this data into a comprehensive report, which provided a balanced view of the employee's performance and helped in identifying areas for development and growth.

Describe your method for tracking tasks assigned to your team.

I use a combination of project management tools and regular communication to track tasks assigned to my team.

Each task is logged into a tool like Jira, where it is assigned a priority level and deadline. I ensure that each team member understands their responsibilities and the timeline. Regular team meetings and one-on-one check-ins help monitor progress and address any issues promptly. For example, while managing a product development team, I used Jira to assign tasks and track their status. This system allowed me to quickly identify any bottlenecks and reallocate resources as needed to keep the project on schedule.

What key metrics were you held to in your last position?

In my last position as a sales manager, I was held to several key metrics, including sales targets, customer acquisition rates, and customer retention rates. Additionally, I was measured on the team's overall performance, such as meeting monthly and quarterly sales goals, the average deal size, and the sales cycle length. For example, we had a quarterly target of increasing sales by 15%. By implementing new sales strategies and providing targeted training to the team, we consistently met or exceeded these targets, contributing significantly to the company's revenue growth.

Initiative

The following questions are designed to assess your initiative:

Describe a situation where you anticipated problems and were able to influence a new direction to avoid those problems.

In my previous role as a project manager, I noticed that our upcoming product launch was at risk due to potential delays from a key supplier.

Anticipating this problem, I proposed a contingency plan that involved sourcing a secondary supplier to ensure we had the necessary components on time. I presented this plan to senior management, highlighting the risks of not having a backup and the benefits of securing our supply chain. They agreed, and we successfully onboarded the secondary supplier.

When the original supplier encountered issues, we seamlessly transitioned to the backup, ensuring the product launch remained on schedule.

How were you assigned tasks in your last role?

In my last role as a marketing coordinator, tasks were assigned based on a combination of project priorities and individual strengths. Our manager would review the project scope and deadlines and then assign tasks during our weekly team meetings. For example, if we had a new campaign, I would typically handle the social media strategy and content creation because of my expertise in those areas. We also used a project management tool like Asana to track assignments and deadlines, ensuring everyone was clear on their responsibilities and the timeline.

What excited you the most about your last position?

The most exciting aspect of my last position as a sales manager was the opportunity to lead a dynamic team and drive significant revenue growth. I enjoyed developing and implementing innovative sales strategies that helped us exceed our targets. One highlight was when we launched a new product line and successfully captured a significant market share within the first six months. Seeing the direct impact of our efforts on the company's bottom line and celebrating those successes with my team was incredibly rewarding.

Can you share a time in school when you exceeded expectations?

In college, I was part of a group project for a business strategy course where we had to develop a comprehensive business plan. I took the lead in coordinating the team's efforts, ensuring that everyone was on the same page and contributing effectively. To exceed expectations, I suggested we conduct additional market research and include a detailed financial projection, which was not required by the professor. Our thorough and well-prepared presentation impressed the professor and our classmates, earning us the highest grade in the class and positive feedback on our innovative approach and detailed analysis.

Interpersonal skills

The following questions are designed to assess your interpersonal skills:

Describe a recent unpopular decision you made and what the outcome was.

In my previous role, I made the unpopular decision to restructure our project timelines to address quality concerns. While the team was initially resistant because it meant longer hours and tighter deadlines, I explained that the changes were necessary to meet client expectations and maintain our reputation.

I implemented weekly check-ins to support the team and ensure we stayed on track. Though it was challenging, the outcome was positive: we delivered a high-quality project that earned praise from the client and led to additional business opportunities.

Can you share an example of the most difficult person you have ever worked with and how you handled your interactions with them?

In a previous role as a sales manager, I worked with a colleague who was consistently negative and resistant to new ideas. To handle this, I tried to understand their perspective and find common ground. I scheduled regular one-on-one meetings to discuss their concerns and provided constructive feedback. During an important project, I assigned tasks that leveraged their strengths and acknowledged their contributions in team meetings. This approach helped build trust and improved our working relationship.

Describe how you contributed to a team environment in your last position.

In my last role as a marketing coordinator, I contributed to a positive team environment by fostering open communication and collaboration. I organized regular brainstorming sessions where everyone could share ideas and feedback freely. For example, during a major product launch, I encouraged team members to bring their unique perspectives to our strategy meetings. I also implemented a peer recognition program to celebrate individual and team achievements.

By creating a supportive and inclusive atmosphere, we not only met our project deadlines but also enhanced team morale and productivity.

Innovation

The following questions are designed to assess your ability to innovate with new ideas to solve problems:

Describe a situation where you had to be innovative at work. What was the outcome?

In my role as a retail manager, we were facing declining foot traffic and sales. I proposed creating an in-store experience event where customers could participate in product demonstrations and workshops. This was a new approach for our store. We partnered with local influencers to promote the event and offered exclusive discounts to attendees.

The event attracted a significant number of customers, resulting in a 25% increase in sales that weekend. The success of this event led to a quarterly schedule of similar events, boosting our overall foot traffic and customer engagement.

Describe a time when you suggested an improvement to your boss. What was the result of the conversation?

At my previous job as a customer service supervisor, I noticed that our response time to customer inquiries was slower than industry standards, leading to customer dissatisfaction. I proposed implementing a live chat feature on our website to provide immediate assistance. My boss was initially skeptical but agreed to pilot the idea. After a three-month trial, our customer satisfaction scores improved by 15%, and live chat became a permanent feature. My boss appreciated the initiative, and it led to further discussions about other potential improvements.

Can you describe a time you used a non-traditional method to solve a problem on the job?

As an event coordinator, I faced a challenge when a key speaker canceled last minute for a major conference. Instead of scrambling to find a replacement, I decided to leverage technology and arranged for the speaker to deliver their presentation via a live video feed. We set up a large screen and ensured the audience could interact with the speaker through real-time Q&A. This non-traditional approach was well received, and attendees appreciated the innovative solution.

The event proceeded smoothly, and feedback indicated that the virtual presentation was just as effective as an in-person one.

When was the last time you had to think "outside of the box"? How did you do it?

While working as a product manager, we were tasked with increasing user engagement for our mobile app. Traditional marketing methods weren't yielding the desired results, so I proposed a gamification strategy. We introduced a points system where users could earn rewards for completing specific actions within the app, like sharing content or inviting friends. This outside-of-the-box approach led to a 40% increase in user engagement within the first two months. It also encouraged viral growth as users were incentivized to invite others, resulting in a significant increase in app downloads.

Describe a time you were required to use creative thinking on the job to solve a problem.

As a logistics coordinator, we encountered a problem where a critical shipment was delayed due to severe weather conditions. To avoid a production halt, I brainstormed alternative solutions and came up with the idea to reroute the shipment through a different logistics provider that could navigate around the affected area. I coordinated with multiple parties and arranged for the goods to be transported by a combination of air and ground transport. This creative problem-solving ensured that the shipment arrived just in time, preventing any disruption in production and demonstrating the value of flexible and innovative thinking.

What is one of your most creative ideas?

In my role as a community outreach coordinator for a non-profit organization, I devised a creative fundraising campaign. The idea was to encourage people to ask for donations to our cause instead of birthday gifts. We provided them with personalized online tools to create fundraising pages and share their stories. This campaign resonated deeply with our supporters and significantly increased our donor base. Over the course of a year, the campaign raised 27% more funds compared to traditional methods and built a stronger, more engaged community around our cause.

Integrity

The following questions are designed to assess your integrity:

Describe a time when you were asked to keep information confidential on the job.

In my previous role as an HR assistant, I was involved in the hiring process for a senior management position. I was entrusted with sensitive information regarding the candidates, including their resumes, interview feedback, and salary expectations.

I ensured that all documents were securely stored and only shared with authorized personnel. I also refrained from discussing any details outside the necessary team. This maintained the integrity of the hiring process and ensured that all candidates' information remained confidential.

Describe a time you acted with integrity.

While working as a sales associate, I discovered an error in a customer's bill that resulted in them being overcharged. Although correcting the mistake meant a loss in commission for me, I immediately informed the customer and rectified the error.

I also reported the issue to my manager and provided suggestions to improve the billing process to prevent future mistakes.

The customer appreciated the honesty and became a loyal client, and my manager acknowledged my integrity, which reinforced the company's reputation for trustworthiness and ethical practices.

Can you tell me about a time when your trustworthiness was challenged and how you reacted?

As a project manager, I was once mistakenly accused of mismanaging a project budget by a colleague. My trustworthiness was challenged, and I responded by calmly providing detailed documentation of all expenditures and project decisions. I organized a meeting with my manager and the colleague to review the records and clarify any misunderstandings. By presenting clear evidence and maintaining transparency, I demonstrated my adherence to the budget and project guidelines. This not only cleared my name but also reinforced my reputation for reliability and accuracy.

Describe a time when you witnessed dishonesty in the workplace and how you handled it.

While working as an administrative assistant, I noticed a co-worker falsifying timesheets to receive overtime pay they hadn't worked. I felt it was important to address this dishonesty, so I discreetly gathered evidence and reported the issue to our HR department. I provided them with the necessary documentation and explained the situation. HR conducted a thorough investigation, and appropriate action was taken to address the misconduct. By handling the situation through the proper channels, I helped maintain a fair and honest work environment.

Describe a time you chose to trust someone on the job and the outcome.

As a marketing team lead, I was tasked with overseeing a critical product launch. I chose to delegate the social media campaign to a new team member who had shown great potential but had not yet handled such a high-profile task.

I provided them with guidance and resources but trusted them to take ownership of the campaign. The team member exceeded my expectations, delivering a creative and effective social media strategy that significantly boosted our product visibility and engagement.

Change management

The following questions are designed to assess your ability to implement change in an organization:

Describe your process for implementing policy changes in your team.

When implementing policy changes, my process involves clear communication, training, and feedback. First, I thoroughly understand the policy change and its implications.

Then, I organize a team meeting to explain the new policy, highlighting the reasons for the change and how it will benefit the team and the organization. I provide written documentation and resources for reference. To ensure everyone is comfortable with the new policy, I conduct training sessions and offer one-on-one support if needed. Finally, I establish a feedback loop to address any concerns and adjust based on team input. This approach ensures that the policy change is understood, accepted, and effectively implemented.

Have you ever experienced resistance when implementing a new idea or policy with your team? If yes, what was your process for dealing with the resistance?

Yes, I experienced resistance when introducing a new project management tool at my previous job. The team was accustomed to the old system and hesitant to adopt the new one. To address this, I first organized a meeting to discuss the benefits of the new tool and how it would streamline our workflow.

I acknowledged their concerns and provided a detailed comparison of the old and new systems. I then arranged hands-on training sessions to help them become familiar with the new tool. I also set up a pilot phase where both systems ran in parallel, allowing the team to gradually transition. By involving the team in the process and providing support, I eased their concerns and successfully implemented the new tool.

Leadership

The following questions are designed to assess your leadership skills:

Describe some ways in which you motivate your co-workers.

I motivate my co-workers by recognizing their achievements, fostering a collaborative environment, and providing opportunities for professional growth. For instance, I regularly organize team meetings to celebrate successes and acknowledge individual contributions. I also encourage open communication and idea sharing, which helps build a sense of camaraderie and teamwork. Additionally, I support my colleagues' career development by providing training resources, mentoring, and opportunities to take on new challenges.

Can you share a time when two members of the team did not work well together and your process for motivating them to work together?

In my previous role as a team lead, I had two team members who frequently clashed over different working styles. To address this, I first spoke with each individually to understand their perspectives and identify the root causes of the conflict.

I then arranged a mediated discussion where they could openly express their concerns and find common ground. I emphasized the importance of teamwork and how their collaboration was essential for the project's success. I also set clear expectations for respectful communication and collaboration. To further support them, I assigned them a joint task where they had to rely on each other's strengths. Over time, their working relationship improved, and they became more effective collaborators.

Describe the toughest situation you have faced when trying to get cooperation from other teams.

The toughest situation I faced was during a company-wide software upgrade that required cooperation from multiple departments, including IT, operations, and sales. Each team had different priorities and concerns, leading to resistance and delays. To overcome this, I organized a cross-functional task force with representatives from each department. I facilitated regular meetings to ensure everyone was informed and had a platform to voice their concerns.

I also worked on aligning the project goals with each department's objectives, highlighting how the upgrade would benefit their specific workflows. By fostering open communication, addressing concerns promptly, and finding common ground, I was able to secure the necessary cooperation and complete the upgrade successfully.

Listening

The following questions are designed to assess your ability to listen to others:

Describe a time when you made a mistake because you did not listen to the advice of others.

In my role as a marketing manager, I once decided to launch a new advertising campaign without fully considering the input from my team, particularly the insights from our market analyst. The analyst had suggested that the target audience's preferences were shifting and recommended a different approach. I was confident in my initial plan and proceeded without incorporating their feedback. As a result, the campaign did not resonate with our audience, leading to lower engagement and sales than expected. Realizing my mistake, I called a meeting to review what went wrong and listened carefully to my team's suggestions. We quickly adjusted our strategy, and the subsequent campaign was much more successful. This experience taught me the importance of valuing and integrating diverse perspectives.

How do you show people that you are listening to them?

I show people that I am listening to them by maintaining eye contact, nodding in agreement, and providing verbal acknowledgments like "*I see*" or "*I understand*." I also ask clarifying questions to ensure I fully grasp their point and repeat back key points to confirm my understanding. For example, during team meetings, I actively take notes and refer to them when discussing action items or follow-ups. Additionally, I make it a point to address any concerns raised by providing thoughtful responses or acting based on their feedback. This approach demonstrates my commitment to listening and valuing their input.

Describe when you think listening is important in your job and when you find it difficult to listen.

Listening is important in my job when gathering requirements for a project or resolving conflicts. For instance, as a project manager, understanding the needs and expectations of clients and team members is essential for successful project delivery.

Active listening helps ensure that everyone is aligned and that any potential issues are addressed early on. However, I sometimes find it difficult to listen during high-pressure situations where multiple tasks demand my attention simultaneously.

In such cases, I remind myself to pause, focus on the person speaking, and prioritize effective communication. By doing so, I can manage my stress and still give the speaker the attention they deserve.

Motivation

The following questions assess your ability to motivate other people and self-motivation:

How do you motivate your team?

I motivate my team by creating a positive and supportive work environment where everyone feels valued and appreciated. I set clear goals and provide regular feedback to help them understand their progress and areas for improvement. Additionally, I celebrate both small and big achievements to keep the team motivated. For example, in my previous role as a sales manager, I introduced a monthly recognition program where top performers were acknowledged and rewarded with incentives like gift cards or extra time off. This not only motivated the team to perform better but also fostered a sense of camaraderie and healthy competition.

Describe your process for assessing when your team members exceed your expectations on projects. How do you reward them?

My process for assessing team members' performance involves setting clear objectives at the start of each project and tracking their progress through regular check-ins. I use **key performance indicators (KPIs)** to measure their achievements objectively. When team members exceed expectations, I recognize their efforts through both formal and informal means.

For example, I might highlight their accomplishments in team meetings or send a personalized thank-you note. Additionally, I recommend them for bonuses or promotions when appropriate. In one instance, a team member's innovative solution to a complex problem significantly improved our project's efficiency. I rewarded them with a bonus and public acknowledgment of their contribution, which boosted their morale and set a positive example for the rest of the team.

Can you share how you get subordinates to work at their peak potential and Produce outcome at a higher level?

To help subordinates work at their peak potential, I focus on providing clear direction, resources, and support. I start by setting challenging yet achievable goals and ensuring that everyone understands their role in achieving them. I offer continuous training and development opportunities to help them enhance their skills. Additionally, I foster an open-door policy where team members feel comfortable discussing any challenges or seeking advice.

For example, during a recent critical project, I conducted regular one-on-one sessions with each team member to address their concerns and provide personalized guidance. This approach not only improved their performance but also boosted their confidence and commitment to the project.

Can you share a time when you positively impacted others?

As a mentor for new hires in my previous company, I had the opportunity to positively impact a junior colleague who was struggling to adapt to the fast-paced environment. I took the time to understand their challenges and provided tailored guidance to help them improve their time management and organizational skills. I also introduced them to other team members and encouraged them to participate in team activities to build their confidence and network. Over time, the colleague became more comfortable and productive, eventually earning recognition for their contributions.

How do you define success in your career?

I define success in my career by the positive impact I have on my team, the projects I lead, and the organization. For me, success is about not only achieving financial targets or completing projects on time but also fostering a collaborative and innovative work environment. It involves helping team members grow and reach their potential while delivering high-quality results that drive the company's objectives.

For example, in my role as a project manager, I consider a project successful if it meets client expectations, stays within budget, and contributes to the professional development of my team. Personal satisfaction and continuous learning are also key components of my career success.

Negotiation

The following questions assess your experience in negotiating:

Can you share the most difficult negotiation you were involved in and the results of that negotiation?

One of the most difficult negotiations I was involved in occurred when I was a purchasing manager at a manufacturing company. We needed to secure a critical component from a supplier who had recently increased their prices significantly. The higher cost would have impacted our product pricing and profitability. I prepared thoroughly by researching market prices, understanding the supplier's position, and identifying alternative suppliers. During the negotiation, I emphasized the long-term business relationship and our volume of orders, proposing a compromise where we would agree to a smaller price increase and a longer-term contract.

After several rounds of discussions, we reached an agreement that met our budget constraints while ensuring the supplier maintained our business, resulting in a mutually beneficial outcome.

Describe a time when you had to bargain with someone. What was your process?

As a sales representative, I had to negotiate a contract with a major client who wanted a significant discount on our services. My process began with understanding the client's needs and constraints. I gathered relevant data, including our cost structure and the client's purchasing history. I also reviewed the value of our services provided to the client. During the negotiation, I presented a detailed proposal highlighting the benefits and value of our services, while also explaining the rationale behind our pricing. I offered a moderate discount in exchange for a longer contract term, which would ensure ongoing business for us.

The client appreciated the transparency and value proposition, and we successfully reached a mutually beneficial agreement.

What do you find to be the most difficult part of negotiation?

The most difficult part of negotiation, in my experience, is finding a balance between meeting the other party's needs and protecting your own interests. It's challenging to ensure that both parties feel they have gained something valuable without compromising too much. For example, during a vendor negotiation, balancing the desire for lower costs with the need to maintain quality and timely delivery can be tricky.

To address this, I focus on building a collaborative atmosphere where both sides openly communicate their priorities and constraints. By understanding each other's positions, we can explore creative solutions that satisfy both parties' essential needs.

Organizational responsibilities

The following questions assess your ability to prioritize organizational responsibilities:

Describe a time when you had to make a difficult choice between your personal and professional life.

While working as a project manager, I faced a situation where a critical project deadline coincided with a significant family event. The project required my full attention due to unexpected technical issues, but the family event was also important to me.

After carefully considering the impact on both sides, I decided to delegate key tasks to my capable team members and set up remote monitoring to stay involved while attending the family event. I communicated openly with my team and ensured they had all the necessary resources and support. This decision allowed me to balance my personal and professional responsibilities effectively, and the project was completed on time without compromising my family commitment.

How do you decide what the top priority is when you schedule your time?

When scheduling my time, I prioritize tasks based on their urgency and importance. I use the **Eisenhower Matrix** to categorize tasks into four quadrants: urgent and important, important but not urgent, urgent but not important, and neither urgent nor important.

For example, if I have a project deadline approaching and a team meeting to prepare for, I focus on completing the project first since it has a clear deadline and significant impact. I also consider the long-term goals and the potential consequences of delaying certain tasks.

This method helps me stay focused on what truly matters and ensures that I address high-priority items efficiently.

Describe what you do when you experience interruptions to your planned schedule.

When I experience interruptions to my planned schedule, I first assess the nature and urgency of the interruption. If it's something that requires immediate attention, I address it promptly while making a mental note of how it affects my schedule. I then adjust my priorities and timeline accordingly.

For example, if an urgent client request comes in during a day planned for internal meetings, I reschedule less critical tasks and delegate where possible to accommodate the urgent request. I also build buffer time into my schedule to account for unforeseen interruptions. This approach helps me stay flexible and maintain productivity despite unexpected changes.

Performance management

The following questions assess your performance management ability:

Describe a time when you helped someone accept change and make the necessary adjustments to move forward. What were the change/transition skills that you used with this person?

In my role as a team leader, our company decided to implement a new software system for project management.

One of my team members was particularly resistant to this change, feeling overwhelmed by the new technology. To help them accept the change and move forward, I first listened to their concerns to understand their apprehensions. I then provided one-on-one training sessions to familiarize them with the software, breaking down the process into manageable steps.

I also highlighted the benefits of the new system, such as improved efficiency and easier collaboration. By using active listening, empathy, and tailored support, I helped the team member transition smoothly and become proficient with the new tool.

Describe how you have empowered others to accomplish tasks.

As a project manager, I believe in empowering my team by giving them ownership of their tasks and providing the necessary support. For example, during a major marketing campaign, I assigned each team member specific responsibilities based on their strengths and interests. I provided them with clear objectives and the resources they needed, but I also gave them the autonomy to approach their tasks creatively. I also held regular check-ins to offer guidance and ensure alignment with the overall project goals.

Describe your process for handling performance reviews.

My process for handling performance reviews is structured yet supportive. I start by gathering data on the employee's performance over the review period, including metrics, feedback from colleagues, and self-assessments. I schedule a dedicated time for the review meeting, ensuring a private and comfortable setting. During the review, I begin with positive feedback to highlight the employee's strengths and achievements. I then discuss areas for improvement, providing specific examples and actionable suggestions. I encourage a two-way dialogue, allowing the employee to share their thoughts and goals. Finally, we collaboratively set objectives for the next period and identify any support or resources they might need.

Can you tell me about a time you had to take disciplinary action against one of the people you supervised?

In my previous role as a retail manager, I had to take disciplinary action against an employee who was consistently late and had received multiple warnings. I followed the company's disciplinary procedures, starting with a private meeting to discuss the issue and understand any underlying reasons for the behavior. Despite offering support and setting clear expectations, the behavior did not improve. I then issued a formal written warning, outlining the consequences of continued tardiness. When the behavior persisted, I made the difficult decision to suspend the employee for a period, making it clear that further infractions could lead to termination.

This action emphasized the importance of punctuality and fairness to the rest of the team, ultimately improving overall discipline.

Give an example of how you provide constructive criticism to your team.

While leading a software development team, I noticed that one of the developers frequently submitted code that required significant revisions. To address this, I scheduled a private meeting to provide constructive criticism. I started by acknowledging their hard work and contributions to the project.

Then, I pointed out specific instances where their code had issues and explained how these affected the overall project timeline. I provided concrete examples and suggested best practices they could follow to improve their coding standards. Additionally, I paired them with a more experienced developer for mentorship. This approach was well received, and over time, the quality of their work improved significantly, benefiting both the individual and the team.

Summary

In this chapter, you learned about common behavioral questions that you might be asked in an interview. Remember, even if you don't have work situations that you can use as examples to answer these questions, you may still have experiences from your personal life that you can use to answer them. In an interview, you will probably only be asked a few of these questions, but it's good to study a few questions from each section so that you are well prepared for interviews.

In the next chapter, the authors will share their final interview advice with you.

Join us on Discord!

Read this book alongside other users. Ask questions, provide solutions to other readers, and much more.

Scan the QR code or visit the link to join the community.

`https://packt.link/SecNet`

16

Final Thoughts

In this chapter, we (the authors) will share our final thoughts and advice on cybersecurity interviews and careers.

The following topics will be covered in this chapter:

- Chris Foulon's final thoughts and advice
- Tia Hopkins' final thoughts and advice
- Ken Underhill's final thoughts and advice

Chris Foulon's final thoughts and advice

As you progress in your cybersecurity career, remember that this milestone signifies the beginning rather than the end. Embrace this achievement as the start of a long journey of learning, contributing to the community, and making the world a safer place. The knowledge gained from this book can aid you in securing the role you aspire to and in making a positive impact on those around you.

Cybersecurity is a continuously evolving field, and we require individuals from diverse backgrounds and with varied perspectives to address the increasing cybersecurity challenges. Share your passion with others, whether it's within your child's PTA group, among friends, or in your local school system. Please assist in preparing the next generation for these challenges.

In writing these final thoughts for you today, I say, you can do it! I received this message from a connection of mine that demonstrated my journey:

"I wouldn't be where I am today without Chris. His guidance and expertise have been crucial in my professional development, offering new insights that have allowed me to excel in my career. Chris's remarkable ability to break down and explain complex problems, paired with his exceptional technical and non-technical skills, has been instrumental to my success. The invaluable information he provides has consistently helped me overcome challenges and seize new opportunities. The Master LLM course he offers has greatly facilitated my learning and navigation of the AI space. Chris not only possesses extensive knowledge but also excels at teaching it, showcasing his deep understanding of these topics. I highly recommend Chris and what he offers; I wouldn't consider calling anyone else."

—*John Grose*

Connect with Chris on LinkedIn: `https://www.linkedin.com/in/christophefoulon`.

Tia Hopkins' final thoughts and advice

You made it to the end of the book! Nice! There's no greater investment than the investments you make in yourself. I'm sure Ken and Chris will give you some incredible tips and words of encouragement to take with you, so I'm going to do my best to keep up with them by offering a few of my personal mottos and beliefs that I hope will help you throughout your cybersecurity career journey.

First, know your *why*. Before you go down the path of investing in education and certification training, go back to step zero and ask yourself, *why am I here?*, *why do I even want to pursue a career in cybersecurity?*, and *am I sure this is where I want to be and not just a career someone told me would be good to pursue?*

Cybersecurity careers can be incredibly rewarding, but I won't pretend for a second that it's easy. Knowing your *why* will continue to guide you, balance you, keep you sane when things get hard (and they absolutely will), and motivate you if you get down on yourself and begin questioning your capabilities. If you don't know why you've chosen this industry, then you're allowing everyone else to define what your career should look like. For example, if you ask someone whether you should pursue a degree or certification, many people will give you an answer based on their opinions and personal experiences. But the right answer is, *it depends*. It depends on *your* desired outcomes and what you want for *your* career – not what worked for someone else in theirs.

My next point... I played tackle football for a number of years and learned a lot about the game and myself. I was at practice one night, going through a linebacker drill, and I was struggling with the technique. I tried over and over again to the point where I got so frustrated that I had to take a break. Finally, I took a deep breath, tried again, and I got it right. I absolutely nailed it! I was so happy. But when I looked over at my coach in search of a smile or some other sign of approval, he wasn't smiling. He didn't nod or even tip his cap. Instead, he looked at me and said, *"Don't just do it till you get it right; do it till you can't get it wrong."* Admittedly, I was incredibly annoyed at the time, but after practice, I thought about it and it made perfect sense. Since that day, I've carried his message with me, and even today, it influences my work ethic. So my message to you is, when you're learning a new skill or concept, don't just practice until you get it right; practice until you can't get it wrong. Trust me when I tell you you'll stand out among your peers and be the go-to person in the room if you adopt this way of thinking. On the path to greatness, good enough should never be good enough.

Finally, I want to leave you with a sort of affirmation I created for myself a while back. Sometimes, we work so hard to achieve our goals that we are too hard on ourselves and push a bit more than we should. I'm guilty of this as well, so I've recently developed the habit of reminding myself to give myself *grace*. From my perspective, G.R.A.C.E. is an acronym for a package deal, and on certain days, I have to embody some of these mindsets more than others, but the total package keeps me grounded. I'll share its contents with you:

- **Grit:**
  - Official definition: The passion and perseverance for long-term and meaningful goals.
  - Tia's translation: *DON'T QUIT.*

- **Resilience:**
  - Official definition: The ability to recover from a setback, adapt well to change, and keep going in the face of adversity.
  - Tia's translation: *ALWAYS GET UP.*

- **Agility:**
  - Official definition: The ability to think and understand quickly.
  - Tia's translation: *BE QUICK ON YOUR FEET.*

- Curiosity:
  - Official definition: A strong desire to learn or know something.
  - Tia's translation: *NEVER STOP LEARNING*.

- Empowerment:
  - Official definition: The knowledge, confidence, means, or ability to do things or make decisions for yourself.
  - Tia's translation: *ASK FOR FORGIVENESS, NOT PERMISSION*.

Your career is your choice and your responsibility. Find your lane, and if you can't find one, create one. Use this book as a tool to help you find and land the job that's right for you – because *you* said so. Good luck!

Connect with Tia on LinkedIn: `https://www.linkedin.com/in/yatiahopkins`.

Ken Underhill's final thoughts and advice

First, congratulations on making it to the end of the book. You have worked hard, and preparing for your job interview is just the beginning of a rewarding career. I have seen some people across social media telling others to *fake it until you make it* for job interviews. My advice is to ignore them because a hiring manager can always see through the lie. Instead, if you don't know the answer to something, just say you don't know but that you know how to find the answer. Remember, nobody in cybersecurity knows every single nook and cranny of cybersecurity, and we all do Google searches from time to time.

Another mistake I see people making is not networking properly. Remember, relationships are a two-way street, so it's important to give to the other person instead of just expecting to take. In the context of using networking to get job interviews, instead of reaching out to someone on LinkedIn and saying, *please, please help me get a job interview*, try instead asking whether they need help with any projects so that you can build your skills, or ask them whether you can help them in any other way. Building real relationships will accelerate your career. In fact, a young woman focused on building a real relationship with me via LinkedIn and Zoom. She had applied to jobs for months, hearing nothing back, but when she connected with me, I was able to connect her with a company that was hiring, and she aced the interview. She now works at a fantastic company with a salary she never dreamed of getting as an entry-level cybersecurity professional and this all happened within two weeks.

Regarding the debate on certifications versus college degrees versus just learning skills on your own, I will say, *it depends*. I'm not a fan of certifications, but I am fully aware that some companies require those or college degrees for jobs. I would suggest reflecting on what career you want and which companies you want to work at, and then deciding which (if any) certifications or college degrees you need. The other thing to keep in mind is that almost everything in a job description is just part of a wish list. If you see that you have a few of the skills listed, then apply for the job. The worst-case scenario is that they say no or you never hear back about an interview. That is totally fine, and you just move on to the next company. Remember, every *no* is one step closer to your *yes*, and sometimes doors close in life because an awesome opportunity is just around the corner for you.

I recommend that you take the latest interview appointment time they have available so that you remain in the interviewer's mind. There is a concept called *recency bias*, which just means recent events in the person's mind are given more importance than past events. By having your interview be the last one of the day, you will be prominent in the interviewer's mind as they make the hiring decision.

I also recommend that you create a short slide presentation for job interviews. This will help you stand out from all the other candidates. The presentation should be about five slides maximum, and the first slide should outline your understanding of why they need someone (you) in this job. You will need to do research on the company and probably speak to some of their employees to get the information for the first slide.

The next three slides should show your 30-, 60-, and 90-day plans for what you are going to do once you are hired to help the organization reach its goals. The final slide should have your name, a tagline, and a short, bulleted list of the key skills that you have for the job. Creating this presentation helps plant the seed in the interviewer's mind that you already have the job and have a plan of action; plus, it gives them a summary in the bulleted list of why you are the most qualified person for the job.

When you reach the end of your job interview, the interviewer typically asks whether you have any questions. I suggest you then ask the interviewer whether you have given them enough information for them to make an offer. If they say no, then ask them what additional information they need from you.

One thing to remember is that the interviewer often works from a checklist during the interview, where they have specific questions they need to ask you and have a pre-determined grading scale to measure you against for the interview.

Here is what an interviewer's checklist might look like:

Cybersecurity Analyst - Competencies					
RATING SCALE: 5 = Excellent 4 = Very Good 3 = Good 2 = Only Fair 1 = Poor					
COMPETENCIES	Minimum Required	RATING			
INTELLECTUAL					
1. Intelligence	5		7. Pragmatism	4	
2. Analysis Skills	4		8. Risk Taking	4	
3. Judgment/Decision Making	4		9. Leading Edge	3	
4. Conceptual Ability	3		10. Education	3	
5. Creativity	4		11. Experience	4	
6. Strategic Skills	3		12. Track Record	5	
PERSONAL					
13. Integrity	5		18. Stress Management	4	
14. Resourcefulness	5		19. Self Awareness	3	
15. Organization / Planning	4		20. Adaptability	4	
16. Excellence	4		21. First Impression	5	
17. Independence	4				
INTERPERSONAL					
22. Likeability	4		27. Communications - Oral	4	
23. Listening	4		28. Communications - Written	4	
24. Customer Focus	5		29. Political Savvy	4	
25. Team Player	5		30. Negotiation	4	
26. Assertiveness	4		31. Persuasion	4	
ADMINISTRATIVE/MANAGEMENT					
32. Technical Competency	4		37. Redeploying B/C Players	4	
33. Mentoring	4		38. Team Building	4	
34. Goal Setting	4		39. Diversity	3	
35. Organizational Accumen	3		40. Running Meetings	3	
36. Accountability	5				

At the time of writing this second edition of the book, it has become **much** more difficult to land a cybersecurity job because of macroeconomic conditions.

Any job opening you see on job aggregation platforms (e.g., LinkedIn, Indeed, Monster, etc.) will literally have over a thousand applications for every job. That's why many of you reading this are not even getting contacted for an interview or even getting a rejection email. Companies simply don't have the resources to look at all these applicants.

Here are some ideas to find your next cybersecurity job:

- **Contact venture capital firms** in your geographic area to see how you might be able to help their portfolio companies using your cybersecurity skills. This can often lead to a job being created for you.

- **Explore the company sponsor lists for large cybersecurity conferences** (e.g., **Black Hat, RSA**, etc.) and look at the lower-tier sponsors. These are typically the startup companies that most people don't know about, and you can often find job openings with less competition. Sometimes, these companies will create jobs for you.

- **Contact your local business community** to see how you can help smaller businesses with cybersecurity. In the United States, most areas have a **Chamber of Commerce, small business administration (SBA)**, startup incubators, or other organizations that work with small businesses. I suggest reaching out to these organizations and asking if you can do a free webinar on a cybersecurity topic. The topic should be relevant to the businesses, which means it's rarely going to be a deeply technical topic. So focus on more general topics like cyber resiliency, security awareness, phishing attacks, website security, etc. Remember, most business owners attending are going to be nontechnical and not care about cybersecurity, so your job is to show them how cybersecurity can make or save them money (or both) in their business. This can help you either get jobs or get contracts with these small businesses.

- Use a search engine or generative AI to **identify the most common questions people ask about your specialty area of cybersecurity**. Another resource to find this information is videos on YouTube. Then, create content (e.g., videos, written posts, etc.) that answers these questions. Doing this helps you become a thought leader in your cyber niche, which then leads to job opportunities.

- **Build a course or write a book** on a cybersecurity area that you know. Remember, you don't need to consider yourself an "expert" because there is always someone out there trying to learn what you already know. Building something like a course shows a hiring manager that you know your topic and that you can be trusted to train other members of the team. That value to the team goes a lot further in getting you hired than a random cybersecurity certification or college degree.

- **Explore non-cybersecurity careers** to get your foot in the door. Many people in cybersecurity you see on social media started their careers doing technical writing, marketing, sales, project management, and other careers working for a cybersecurity product or service company. Then, after a few years working there, they made the transition to the job they really wanted. This is by far the easiest approach to getting a job and a cybersecurity company listed on your résumé.

Once you find a job and start applying, be sure to use a cover letter. The advantage of a cover letter is that it gives you additional space to explain why you are interested in a job and why you are the best fit. Think of this as a sales page on a website that tells you why you should buy something and the benefits you will get if you do. In addition, recruiters will search your résumé and cover letter for keywords and whether your background is relevant for a position, so a cover letter gives you additional space to help them make this connection.

The final thought I will leave you with is my technique for passing exams and interviews over the years. I would simply take a piece of paper and a pen and write out exactly what I wanted to happen during the exam or interview as if it had already happened. This may seem too simple to work, but it has always worked for me.

As an example, if I wanted a job as a SOC Analyst, I would write, *"The interview for the SOC Analyst job went great. They didn't ask me any difficult questions and it felt more like a conversation with friends than it did an interview. I was excited to receive their offer within a week after the interview."*

The people who consistently struggle to find their first job in cybersecurity do not follow this manifestation approach and the other information shared in this book. You can be like them and try applying to thousands of jobs and hope to hear back, or you can take the road less traveled and apply the knowledge you have gained in this book to help you get your dream career faster.

The choice is yours.

Connect with Ken on LinkedIn: `https://www.linkedin.com/in/kenunderhill`.

Summary

Congratulations! You made it to the end of this book, but not the end of your career journey. In this book, you learned about some common interview questions for a variety of cybersecurity careers, the most common behavioral interview questions that are asked, and how you can *hack* yourself.

Join us on Discord!

Read this book alongside other users. Ask questions, provide solutions to other readers, and much more.

Scan the QR code or visit the link to join the community.

`https://packt.link/SecNet`

packt.com

Subscribe to our online digital library for full access to over 7,000 books and videos, as well as industry leading tools to help you plan your personal development and advance your career. For more information, please visit our website.

Why subscribe?

- Spend less time learning and more time coding with practical eBooks and Videos from over 4,000 industry professionals

- Improve your learning with Skill Plans built especially for you

- Get a free eBook or video every month

- Fully searchable for easy access to vital information

- Copy and paste, print, and bookmark content

At www.packt.com, you can also read a collection of free technical articles, sign up for a range of free newsletters, and receive exclusive discounts and offers on Packt books and eBooks.

Other Books You May Enjoy

If you enjoyed this book, you may be interested in these other books by Packt:

The Ultimate Kali Linux Book - Third Edition: Harness Nmap, Metasploit, Aircrack-ng, and Empire for cutting-edge pentesting

Glen D Singh

ISBN: 9781835085806

- Establish a firm foundation in ethical hacking
- Install and configure Kali Linux 2024.1
- Build a penetration testing lab environment and perform vulnerability assessments

- Understand the various approaches a Penetration Tester can undertake for an assessment
- Gathering information from Open Source Intelligence (OSINT) data sources
- Use Nmap to discover security weakness on a target system on a network
- Implement advanced wireless pentesting techniques
- Become well-versed with exploiting vulnerable web applications

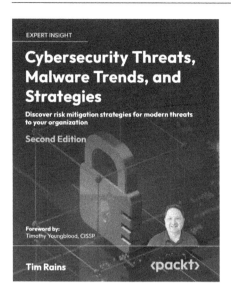

Cybersecurity Threats, Malware Trends, and Strategies - Second Edition: Discover risk mitigation strategies for modern threats to your organization

Tim Rains

ISBN: 9781804613672

- Discover enterprise cybersecurity strategies and the ingredients critical to their success

- Improve vulnerability management by reducing risks and costs for your organization

- Mitigate internet-based threats such as drive-by download attacks and malware distribution sites

- Learn the roles that governments play in cybersecurity and how to mitigate government access to data

- Weigh the pros and cons of popular cybersecurity strategies such as Zero Trust, the Intrusion Kill Chain, and others

- Implement and then measure the outcome of a cybersecurity strategy

- Discover how the cloud can provide better security and compliance capabilities than on-premises IT environments

Packt is searching for authors like you

If you're interested in becoming an author for Packt, please visit `authors.packtpub.com` and apply today. We have worked with thousands of developers and tech professionals, just like you, to help them share their insight with the global tech community. You can make a general application, apply for a specific hot topic that we are recruiting an author for, or submit your own idea.

Share your thoughts

Now you've finished *Hack the Cybersecurity Interview*, we'd love to hear your thoughts! Scan the QR code below to go straight to the Amazon review page for this book and share your feedback or leave a review on the site that you purchased it from.

`https://packt.link/r/1835461298`

Your review is important to us and the tech community and will help us make sure we're delivering excellent quality content.

Index

C

Download a free PDF copy of this book

Thanks for purchasing this book!

Do you like to read on the go but are unable to carry your print books everywhere?

Is your eBook purchase not compatible with the device of your choice?

Don't worry, now with every Packt book you get a DRM-free PDF version of that book at no cost.

Read anywhere, any place, on any device. Search, copy, and paste code from your favorite technical books directly into your application.

The perks don't stop there, you can get exclusive access to discounts, newsletters, and great free content in your inbox daily.

Follow these simple steps to get the benefits:

1. Scan the QR code or visit the link below:

https://packt.link/free-ebook/9781835461297

2. Submit your proof of purchase.
3. That's it! We'll send your free PDF and other benefits to your email directly.